THE WONDERS OF PHYSICS

THE WONDERS OF PHYSICS

L. G. Aslamazov
Late Professor, Moscow Technological University

A. A. Varlamov
Moscow Technological University

Scientific Editor
A. A. Abrikosov Jr

Translators
A. A. Abrikosov Jr & D. Znamenski

World Scientific
Singapore • New Jersey • London • Hong Kong

Published by

World Scientific Publishing Co. Pte. Ltd.

P O Box 128, Farrer Road, Singapore 912805

USA office: Suite 1B, 1060 Main Street, River Edge, NJ 07661

UK office: 57 Shelton Street, Covent Garden, London WC2H 9HE

British Library Cataloguing-in-Publication Data
A catalogue record for this book is available from the British Library.

THE WONDERS OF PHYSICS

ISBN 981-02-4346-4

Printed in Singapore by World Scientific Printers

To our teachers and friends...

Preface

Author's preface to the English Edition

It is my great pleasure to see the book, written together with Lev Aslamazov, to appear in English. The original Russian edition, followed by the Italian one, were accepted enthusiastically by readers and I hope that the wide English-speaking audience will find the book to merit attention too. It would be a fair tribute to the good memory of my friend and coauthor.

The reasons which pushed us to write this book were curiousity and our wish to share with others the admiration for the beauty of Physics in its manifestations in the Nature. The authors have devoted a lot of time to physics teaching of students of various levels, from gifted beginners to mature PhD students. All this experience has convinced us that, besides the evident necessity of regular and careful study of the discipline, an "artistic" approach, in which the teacher (or the author) proves the importance of Physics in habitual everyday phenomena, is vital. I hope that we succeeded to pass to the reader our feeling of Physics not only on the cover but also in the text of the book.

I would like to express my deep gratitude to many friends and colleagues, without whom this edition would not appear. In first turn this is my old and dear friend Dr. Alex Abrikosov (Jr.), whose enthusiasm, thorough scientific care and translation gave birth of the English version. His contrubution to the project was considerably enforced by the collaboration of my other friend and our common alumnus Dr. Dmitriy Znamenski, who has become almost a native speaker of English last years.

I would like to acknowledge the contribution of my coauthours and friends Professor A. Buzdin, Dr. C. Camerlingo, Dr. A. Malyarovski and my old teacher of Physics Dr. A. Shapiro. Several chapters of this book were written basing on our mutual publications in different journals.

Special thanks are addressed to my friend Professor A. Rigamonti, whose encyclopaedic erudition and enthusiasm permitted to realize the Italian edition of the book and considerably adorned it.

Finally I would like to thank warmly my Italian and Russian editors: Dr. D. De Bona, Dr. T. Petrova, Dr. V. Tikhomirova and Dr. L. Panyushkina without whose professionalism and collaboration in preraration of previous editions the present one would not appear.

In conclusion I would like to cordially mention on behalf of mine and Alex Abrikosov (Jr.) three more people. Two of them are Alex's parents and our teachers of Physics and life, Alexei and Tatyana Abrikosov. The third one is our common friend from University times Serguei Pokrovski. These people played the foremost part in our formation.

A. A. Varlamov,

(Rome, 2000).

From the foreword to the Russian edition

The science of physics was at the head of scientific and technical revolution of the twentieth century. Nowadays successes of physics continue to determine the direction of forthcoming progress of the humanity. The bright example of that is the recent discovery of the high-temperature superconductivity which may quite soon radically change the entire edifice of modern technology.

However, delving deeper into the mysteries of the macrocosm and microparticles, scientists move further and further away form the traditional school physics with its transformers and bodies, thrown at an angle to the horizontal, namely, from what most of the people believe to be physics. The goal of popular literature is to bridge the gap, to open to curious readers the excellence of modern physics and to demonstrate its major achievements. The difficult task that does not tolerate dabbling.

The book in your hands develops the best traditions of this kind of literature. Written by working theoretical physicists and, in the same time, the dedicated popularizers of scientific knowledge, clear and captivating in manner, it brings the reader to the latest achievements of the quantum solid-state physics; but on the way it shows how laws of physics reveal themselves even in trivial, at first glance, episodes and natural phenomena around us. And what is most important, it shows the world with the eyes of scientists, capable to "prove the harmony by algebra".

It was a great loss that one of the authors of the book, the well-known specialist in the theory of superconductivity, professor L. G. Aslamazov, who for a long time was the vice-editor of the *"Quantum"* popular journal, did not live till the book coming out.

I hope that the most different readers, ranging from high-school students to professional physicists, will find this book, marked by its extremely vast scope of encompassed questions, a real interesting, enjoyable and rewarding reading.

Academician A. A. Abrikosov,

(Moscow, 1987).

Translator's note

The offer to translate this book into English was a great honor for me. Now I'm your interpreter in the marvelous land of physics. But this is not a simple coincidence.

First of all, for me physics is a sort of "family business" that you, no doubt, might have guessed. Many of people, whom I remember warmly from my first days, afterwards turned to be physicists. As a ten years old schoolboy I remember (then postgraduate, later professor) Lev Aslamazov sunbathing on the Odessa beach[a], then, in the high-school, I met my best friend Andrei Varlamov. We made our decision and both entered the Moscow Physical-Technical Institute. For long hours we discussed and argued about many things related and not related to physics. Some of the topics in this book awake remembrances of those days.

Not the last role in this "physical orientation" belonged to the newly established in Moscow by the enthusiastic young team popular journal *"Kvant"*. (Its English translation is known now as *"Quantum"*.) L. Asla-

[a]Odessa is the city on the Black Sea coast where traditional spring symposia on theoretical physics were held.

mazov was at the very origin of it and his article *"Meandering down to the sea"* appeared in the first issue. Getting older we started writing ourselves. And almost every chapter of this book once has appeared under the *"Kvant's"* cover.

Somewhere among papers I keep the draft of my first popular paper. Leva (as everybody called him) rejected it, explaining, that we must not simply write about what we knew from textbooks, but find new bright and clear illustrations to our knowledge. According to him, this was the main and most difficult task in popularization of science. And, as you shall see, this is the spirit of the present book.

The love to physics swung the balance in favor of translating the book being not a native English speaker. I hope that a share of nonlinguistic knowledge that I tried to invest in the text will, at least partially, compensate its "Russian flavor", and you will rather be amused than annoyed by some inevitable slips.

Sure enough I would not take the risk alone and what you read is a result of real collaboration with my fellow translator Dmitriy Znamensky from whom I learned so much. Writing this note myself is only the privilege of the old acquaintance and this must by no means belittle his contribution. You may feel his vivid style yourself when reading Chapters 8–12, 14–16 and Chapter 21.

But to the work on translation we tried to commemorate the great scientists of the past and supplied the text with short biographical footnotes.

A. A. Abrikosov, jr., (—*A. A.*).

(Moscow, 2000).

Contents

PART I

Outdoor physics

From the first part of the book the reader will learn why rivers are winding and how they wash their banks out, why the sky is blue and the white horses are white. We are going to tell you about properties of the ocean, about winds and the role of the Earth's rotation.

In a word we shall present examples of how laws of physics work on a world scale.

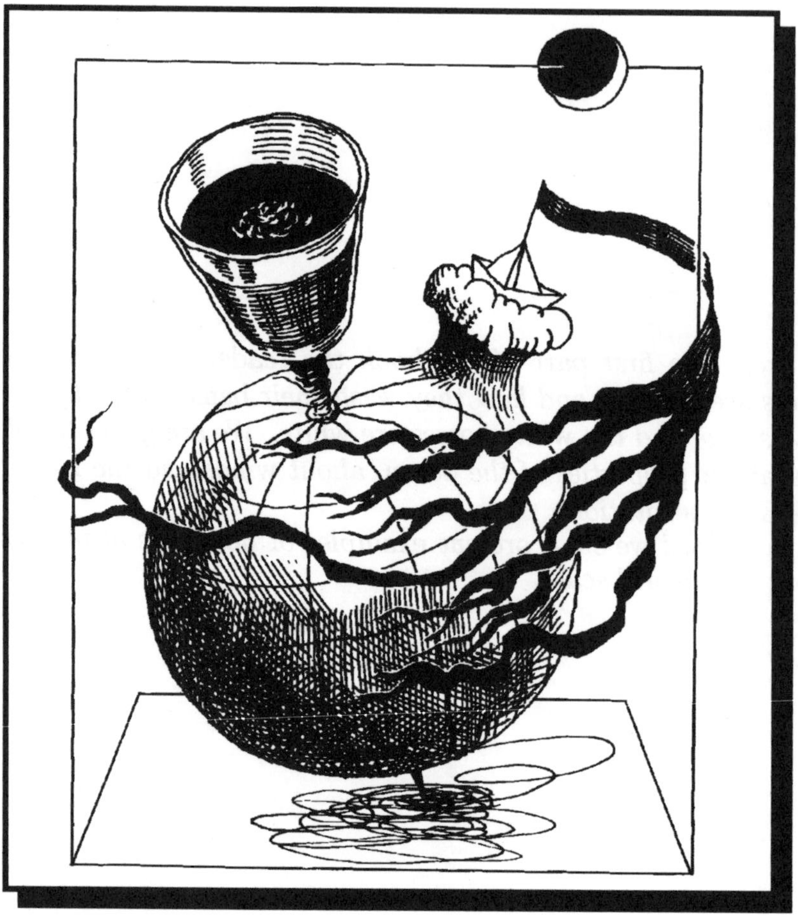

Chapter 1

Meandering down to the sea

Have you ever seen a straight river without bends? Of course a short section of a river may cut straight but no rivers have no bends at all. Even if the river flows through a plain it usually loops around and the bends often repeat periodically. Moreover, as a rule one bank at the bend is steep while the other slopes gently. How could one explain these peculiarities of river behavior?

Hydrodynamics is the branch of physics that deals with the motion of liquids. Although now it's a well-developed science, rivers are too complicated natural objects and even hydrodynamics can't explain every feature of behavior. Nevertheless, it can answer many questions.

You may be surprised to learn that even great Albert Einstein[a] gave time to the problem of meanders. In the report delivered in 1926 at a meeting of the Prussian Academy of Sciences, he compared the motion of river water to swirling of water in a glass. The analogy allowed him to explain why rivers choose the twisted paths.

Let's try to understand this too, at least qualitatively. And let's start with a glass of tea.

1.1 Tea-leaves in a glass

Make a glass of tea with loose tea-leaves (no tea-bags!), stir it well, and take the spoon off. The brew will gradually stop and the tea-leaves will gather

[a] A. Einstein, (1879-1955), German physicist, US citizen from 1940; creator of the theory of relativity; Nobel Prize 1921.

in the center of the bottom. What made them come there? To answer this question let us first determine what shape takes the surface as the liquid swirls in the glass.

The tea-cup-experiment shows that the surface — in our case of the tea — gets curved. The reason is clear. In order to make particles of water move circularly, the net force acting on each of them must provide a centripetal acceleration. Consider a cube of situated in the liquid at a distance r from the axis of rotation, Fig. 1.1, *a*. Let the mass of tea in it be Δm. If the angular speed of rotation is ω then the centripetal acceleration of the cube is $\omega^2 r$. It comes as the result of the difference of the pressures acting onto the faces of the cube (the left and right faces in Fig. 1.1 *a*). So,

$$m\,\omega^2\,r = F_1 - F_2 = (P_1 - P_2)\,\Delta S, \tag{1.1}$$

where ΔS is the area of the face. The pressures P_1 and P_2 are determined by the distances h_1 and h_2 from the surface of the liquid:

$$P_1 = \rho\,g\,h_1 \qquad \text{and} \qquad P_2 = \rho\,g\,h_2, \tag{1.2}$$

where ρ is the density of the liquid and g is the free fall acceleration. As soon as the force F_1 must be greater than F_2, so h_1 must exceed h_2 and the surface of the rotating liquid must be curved, as shown in Fig. 1.1. The faster the rotation is, the greater is the curvature of the surface.

One can find the shape of the curved surface of the revolving liquid. It turns out to be a paraboloid — that is, a surface with a parabolic cross section[b].

As long as we continue stirring the tea with the spoon, we keep it swirling. But when we remove the spoon the viscous friction between layers of the liquid and the friction against the walls and bottom of the glass will convert the kinetic energy of liquid into heat, and the motion will gradually come to rest.

As the rotation slows down, the surface of the liquid flattens. In the mean time vortex currents directed as shown in Fig. 1.1, *b* appear in the liquid. The vortex currents are formed because of the nonuniform deceleration of the liquid at the bottom of the glass and at the surface. Near the bottom, where the friction is stronger, the liquid slows down more effectively than at the surface. So, despite being at equal distances from the

[b]The form of the surface is parabolic only if the liquid is rotated together with the glass as a whole. This is called ***rigid rotation***. —*A. A.*

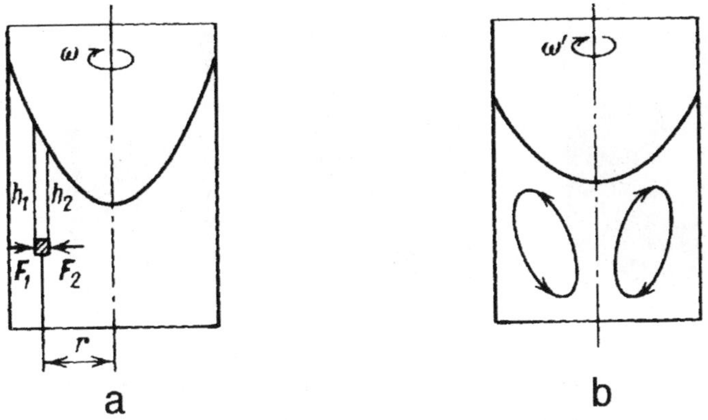

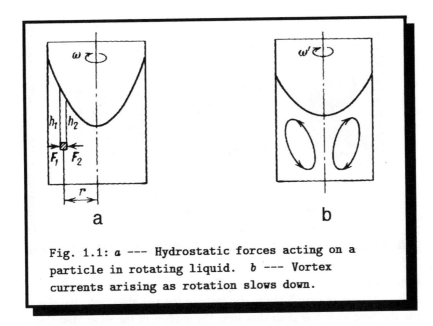

Fig. 1.1: *a* --- Hydrostatic forces acting on a particle in rotating liquid. *b* --- Vortex currents arising as rotation slows down.

axis of rotation particles of liquid acquire different speeds: the ones that are closer to the bottom become slower than those near the surface. However the net force due to the pressure differences is the same for all these particles. This force can't cause the required centripetal acceleration of all the particles at once (as in was in the case of the uniform rotation with the same angular speed). Near the surface the angular speed is too large, and particles of water are thrown to the sides of the glass; near the bottom the angular speed is too low, and the resultant force makes water move to the center.

Now it is clear why tea-leaves gather in the middle of the bottom, Fig. (1.2). They are drawn there by vortex currents that arise due to the nonuniform deceleration. Of course, our analysis is simplified but it accurately grasps the main points.

1.2 How river-beds change

Let's consider the motion of water at a river bend. The picture lively resembles what we have observed in our glass of tea. The surface of water inclines inside the bend in order that pressure differences produced the

Fig. 1.2: The tea-cup experiment. Vortex currents drive tea-leaves to the center of the bottom.

necessary centripetal accelerations. (Figure 1.3 shows schematically a cross section of bending river.) Quite similarly to the tea glass, velocity of water near the bottom is less than that near the surface of the river (distribution of velocities with depth is shown by vectors in Fig. 1.3). Near the surface the net difference of hydrostatic pressures can't make the faster water particles follow the curve and the water is "thrown" to the outer shore (away from the center of the bend). Near the bottom, on the other hand, the velocity is small, so the water moves toward the inner shore (to the center of the bend). Hence additional circulation of water appears in addition to the main flow. The figure 1.3 shows the direction of water circulation in the transverse plane.

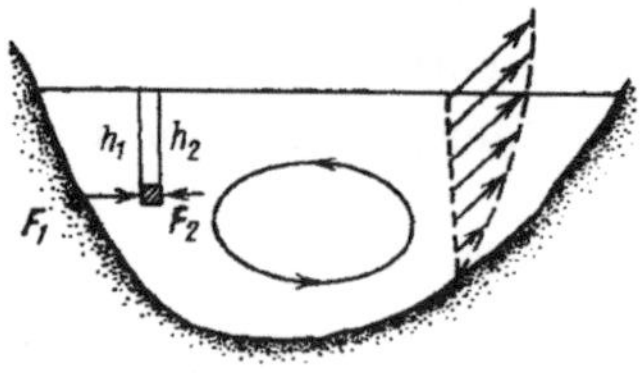

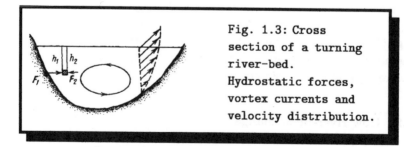

Fig. 1.3: Cross section of a turning river-bed. Hydrostatic forces, vortex currents and velocity distribution.

The circulation of water causes soil erosion. As a result, the outer bank is undermined and washed out while the soil gradually settles along the inner shore, forming an ever thickening layer (remember the tea-leaves in the glass!). The shape of the river-bed changes so that the cross section resembles that shown in Fig. 1.4. It's also interesting to observe how the velocity of the stream varies across the river (from bank to bank). In straight stretches water runs most quickly in the middle of the river. At bends the line of fastest flow shifts outwards. This happens because it's more difficult to turn fast-moving water particles than slow-moving ones. A larger centripetal acceleration is necessary. But the greater is velocity of the flow, the greater is the circulation of the water, and consequently the soil erosion. That's why the fastest place in a river-bed is usually the deepest one — the fact well known by river pilots.

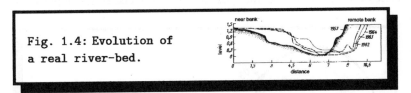

Fig. 1.4: Evolution of a real river-bed.

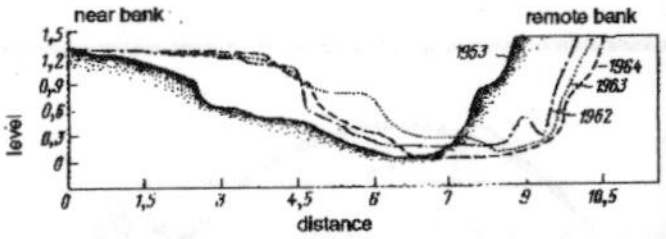

Soil erosion along the outer bank and sedimentation along the inner one result in gradual shift of the entire river-bed away from the center of the bend increasing thereby the river meander. Figure 1.4 shows the very same crosssection of a real river-bed at several years intervals. You can clearly notice the shift of the river-bed and the increase of its meander.

So even an occasional slight river bend — created, for example, by a landslide or by a fallen tree — will grow. Straight flow of a river across a plain is unstable.

1.3 How meanders are formed

The shape of a river-bed is largely determined by the relief of the terrain it crosses. A river passing a hilly landscape winds in order to avoid heights and follows valleys. It "looks for" a path with the maximum slope.

But how do rivers flow in open country? How does the described above instability of straight river-bed with respect to bending influence the course of a stream? The instability must increase the length of the path and make the river wind. It's natural to think that in the ideal case (an absolutely flat, homogeneous terrain), a periodic curve must appear. What will it look like?

Geologists have put forth the idea that at their turns, paths of rivers flowing through plains should take the form of a bent ruler.

Take a steel ruler and bring its ends together. The ruler will bend like it is shown in Fig. 1.5. This special form of elastic curve is called the ***Euler curve*** after the great mathematician Leonhard Euler[c] who has analyzed it theoretically. The shape of the bent ruler has a wonderful property: of all possible curves of a fixed length connecting two given points, it has the minimum average curvature. If we measure the angular deflection θ_k, Fig. 1.5, at equal intervals along the curve and add up their squares then the sum $\theta_1^2 + \theta_2^2 + \ldots$ will be minimal for the Euler curve. This "economic" feature of the Euler curve was basic for the river-bed shape hypothesis.

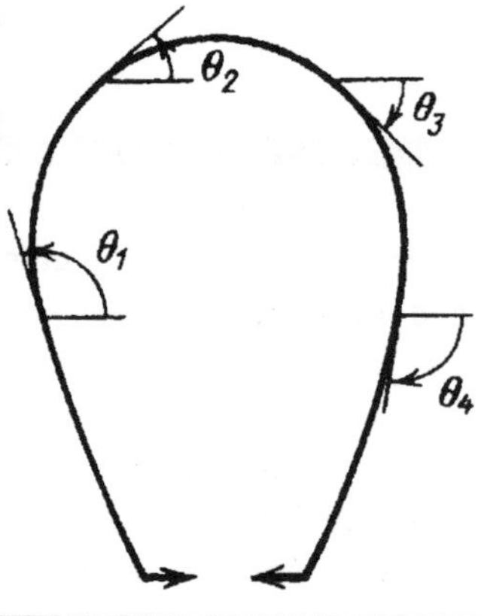

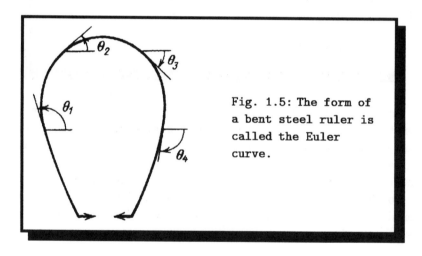

Fig. 1.5: The form of a bent steel ruler is called the Euler curve.

To test this hypothesis geologists modeled a changing river-bed. They passed water through an artificial channel in light erosible homogeneous medium composed of small weakly held together particles. Soon the straight channel began to wander, and the shape of the bend was described by the Euler curve (Fig. 1.6). Of course, nobody has ever seen such a perfect river-bed in nature (because of the heterogeneity of the soil, for instance). But rivers flowing through plains usually do meander and form periodic

[c]L .Euler, (1707–1783), Swiss-born mathematician and physicist; member of Berlin, Paris, St. Petersburg academies and of the London Royal Society; worked a long time in Russia.

structures. In Fig. 1.6 you can see a real river-bed and the Euler curve (the dashed line) that approximates its shape best of all.

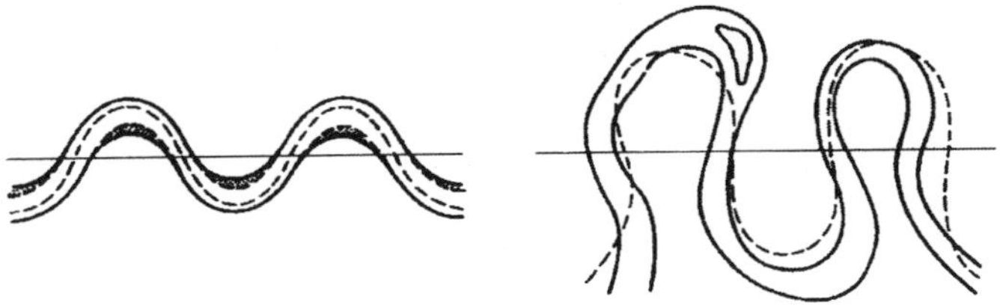

Fig. 1.6: *a*-- Modelling meanders in laboratory. Channel-bed develops periodic Euler bends (the dashed line). *b*-- A real river and the nearest Euler curve (the dashed line).

By the way, the word "meander" itself is of ancient origin. It comes from the *Meander*, a river in Turkey famous for its twists and turns (now called the *Menderes*). Periodic deflections of ocean currents and of brooks that form on surfaces of glaciers are also called meanders. In each of these cases, random processes in a homogeneous medium give rise to periodic structures; and though the reasons that bring meanders about may differ, the shape of resulting periodic curves is always the same.

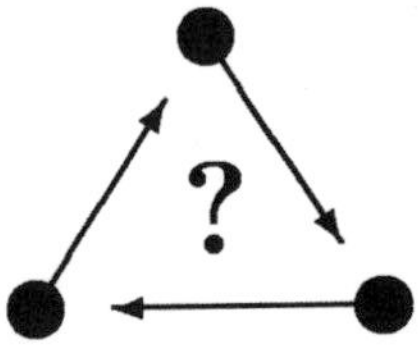

Show that the surface of uniformly (rigidly) revolving liquid takes the parabolic form.

Chapter 2
Rivers from lakes

...Old Baikal had more than three hundred sons and the sole daughter, beautiful Angara.

The ancient legend.

Let sceptical reader disbelieve the epigraph and search geographical atlases. Jokes apart, he will discover that 336 rivers fall into the lake of Baikal and only one, the mighty Angara, has it for the source. It turns out that Baikal is not the exception. No matter how many rivers fall into a lake only one comes out of it as a rule.

For example many rivers run to the Ladoga lake but only Neva escapes it; Svir is the only outflow of the Onega lake *etc.* This fact may be explained. The outcoming water prefers the deepest river-bed and other possible exits are left above the level of the lake. It is hardly probable that openings of several would-be river-beds have the same elevation. In case of the copious water supply the brimming lake can send forth two streams. However such a situation is not steady and may take place only in relatively young (recently formed) lakes. Little-by-little the deeper and faster stream will wash away the bed increasing the outflow. As a result the level of the lake will decline and the weaker flow will be gradually silted. Thus only the deepest of the outcoming rivers will survive.

In order that a lake was the source of two rivers it is necessary that their origins lay exactly at the same level. This case is called a ***bifurcation*** (the term is now widely used by mathematicians to indicate the doubling of the number of solutions of an equation). Bifurcations are uncommon and usually only a single river comes out of a lake.

The same laws may be applied to rivers. It is well known that rivers readily flow together, whereas forks are relatively infrequent. Streams always prefer the steepest descending curve. The probability of splitting of this curve is small. Nevertheless the situation changes at the river delta where the main stream divides into many smaller channels.

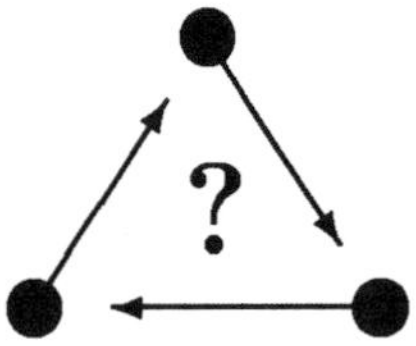

Try to figure out why river behaves so strange when approaching big lake or sea.

Chapter 3

The oceanic phone booth

The walls, indeed, had ears. Or, rather, an ear. A round hole with a tube — a sort of secret telephone — conveyed every word said in the dungeon right to the chamber of Signore Tomato.

Gianni Rodari, The adventures of Cipollino.

Not so long ago — in mid-forties, to put a number on it — scientists from the USSR and US discovered an amazing phenomenon. Sound waves propagating in the ocean could sometimes be detected thousands of kilometers away from the source. In one of the most successful experiments, the sound from an underground explosion set off by scientists on the coast of Australia traveled halfway around the globe and was registered by another group of researchers in Bermuda, some 19.600 km away (a record distance for the propagation of pulse sound signals). This means that the intensity of the sound didn't change greatly as it ran away from the source. What is the mechanism for such long-distance propagation of sound?

It looked as if the ocean contained an acoustic waveguide — that is, a channel along which sound waves traveled practically without attenuation (loss of strength). You have read about such a means in the epigraph.

Another example of acoustic waveguide is the tube used on ships from time immemorial. The ship's captain uses a tube to give orders to the engine room from the bridge. It's interesting that the attenuation of sound traveling in air along a waveguide is so small that if we constructed a tube 750 km long, it could serve as a "telephone" for calls between Pittsburgh and Detroit. But it would be inconvenient to chat over such a line, because your friend at the other end would have to wait a half-hour to hear your words.

We should emphasize that the reflection of a wave from a waveguide's boundaries is a crucial feature of the waveguide: it's because of this very property that the wave energy doesn't scatter in all directions but follows the given one.

These examples would lead us to suppose that the propagation of sound over extremely large distances in the ocean is due to some sort of waveguide mechanism. But how is such a gigantic waveguide formed? Under what conditions does it appear, and what are the reflective boundaries that make the sound waves to travel so far?

Since the ocean surface can reflect sound fairly well, it might serve the upper boundary of the waveguide. The ratio of the intensity of reflected wave to that of a wave penetrating the interface between two media strongly depends on the densities and the speeds of sound in them. If the media differ substantially then even the sound falling normally onto a flat interface will be practically completely reflected. The densities of air and water differ a thousand times, the ratio of sound velocities in them is 4.5. Therefore the intensity of the normal wave passing to air from water is only 0.01 % of the intensity of the incident sound. The reflection is still stronger when the wave falls onto the interface obliquely. But, of course, the ocean surface can't be perfectly flat because of the ever-present waves. This causes chaotic reflections of sound waves and disturbs the waveguide character of their propagation.

The results aren't any better when the sound waves reflect at the ocean floor. The density of sediments at the bottom of the sea is usually within the range 1.24–2.0 g/cm^3, and the velocity of sound propagation in these sediments is only 2–3% less than that in water. So when a sound wave hits the bottom a significant amount of its energy is absorbed. .

Once the ocean floor poorly reflects sound it can't serve the lower boundary of the waveguide.

The boundaries of the oceanic waveguide must be sought somewhere in between the floor and the surface. And that's where they were found. The boundaries turned out to be the water layers at certain depths in the ocean.

How do sound waves reflect from the "walls" of the oceanic acoustic waveguide? To answer this question we'll have to examine the mechanism of sound propagation in the ocean.

3.1 Sound in water

Up to now, as we've talked about waveguides, our unspoken assumption was that the speed of sound in them is constant. But the speed of sound in the ocean varies from $1.450\,m/s$ to $1.540\,m/s$. It depends on temperature, salinity, hydrostatic pressure and other factors. The increase in hydrostatic pressure $P(z)$ with depth z, for instance, adds to the speed of sound $1.6\,m/s$ per 100 meters down. An increase in temperature also adds to the speed of sound. However, the water temperature, as a rule, falls rapidly as one descends from the well-warmed upper layers to the ocean depths, where the temperature is practically constant. Due to the interplay of these two mechanisms — the hydrostatic pressure and the temperature — the dependence of the sound velocity $c(z)$ on depth in the ocean looks like that shown in Fig. 3.1. Near the surface the rapidly dropping temperature takes the upper hand. Here the speed of sound decreases with depth. As we plunge deeper, the rate of decrease in temperature slows, but the hydrostatic pressure continues to grow. At some depth the two factors balance: the speed of sound reaches its minimum. Deeper down the sound velocity starts to grow due to the rise in hydrostatic pressure.

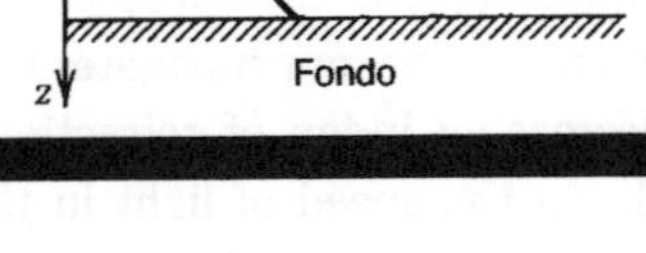

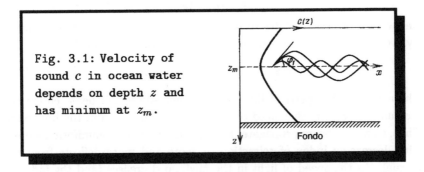

Fig. 3.1: Velocity of sound c in ocean water depends on depth z and has minimum at z_m.

We see that the speed of sound in the ocean depends on depth and this influences the nature of the sound propagation. To understand how "sound beams" travel in the ocean, we'll turn to an optical analogy. We'll examine how a light beam propagates in a stack of flat parallel plates with varying indices of refraction. Then we'll generalize our findings to a medium with smoothly varying refraction index.

3.2 Light in water

Let's consider a pile of flat parallel plates with varying indices of refraction $n_0, n_1, \dots n_k$, where $n_0 < n_1 < \dots < n_k$ (see Fig. 3.2). Assume that the light beam falls onto the upper plate at an angle α_0 relative to the normal. After refraction it leaves the 0–1 boundary at an angle α_1 and that is the incidence angle for the 1–2 boundary. Upon refracting at the next interface the beam hits the 2–3 boundary at an angle α_2 and so on. According to Snell's[a] law, we have:

$$\frac{\sin\alpha_0}{\sin\alpha_1} = \frac{n_1}{n_0}, \quad \frac{\sin\alpha_1}{\sin\alpha_2} = \frac{n_2}{n_1}, \dots \quad \frac{\sin\alpha_{k-1}}{\sin\alpha_k} = \frac{n_k}{n_{k-1}}.$$

Remembering that the ratio of the refraction indices of two media is the inverse of the ratio of the speeds of light in these media, we may rewrite these equations in the following form:

$$\frac{\sin\alpha_0}{\sin\alpha_1} = \frac{c_0}{c_1}, \quad \frac{\sin\alpha_1}{\sin\alpha_2} = \frac{c_1}{c_2}, \dots \quad \frac{\sin\alpha_{k-1}}{\sin\alpha_k} = \frac{c_{k-1}}{c_k}.$$

Multiplying the series of equations by one another, we get

$$\frac{\sin\alpha_0}{\sin\alpha_k} = \frac{c_0}{c_k}.$$

Reducing thickness of plates to zero while increasing their number to infinity, we'll approach the generalized law of refraction (the Snell's law):

$$c(z)\,\sin\alpha(0) = c(0)\,\sin\alpha(z). \tag{3.1}$$

where $c(0)$ is the speed of light at the point where the beam enters the medium and $c(z)$ is the speed of light at a distance z from the boundary. Thus, as a light beam propagates through an optically nonuniform medium with decreasing index of refraction, it more and more deflects from the normal. As the speed of light in the medium decreases (and the index of refraction increases) the beam gradually turns parallel to the interface.

If we know how the speed of light varies in an optically nonuniform medium, we can use the Snell's law to find the trajectory of any beam in it. Sound beams propagating in an acoustically nonuniform medium, where the speed of sound varies, get deflected exactly in the same way. An example of such a medium is the ocean.

[a]W. Snell van Royen, (died 1626), Dutch mathematician.

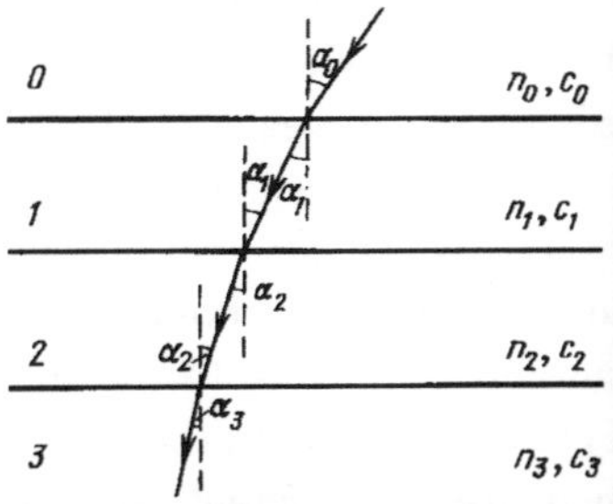

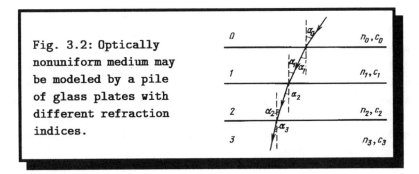

Fig. 3.2: Optically nonuniform medium may be modeled by a pile of glass plates with different refraction indices.

3.3 Water waveguides

Now let's get back to the question of sound propagation in the oceanic acoustic waveguide. Imagine that the sound source is located at the depth z_m corresponding to the minimum sound velocity (Fig. 3.3). How do the sound beams travel as they leave the source? The beam propagating along a horizontal line is straight. But the beams leaving the source at an angle with the horizontal will bend because of sonic refraction. Since the speed of sound increases both up and down from the level z_m, sound beams will bend towards the horizontal. At a certain point the beam will get parallel to the horizontal and after being reflected it will turn back toward the line $z = z_m$, Fig. 3.3.

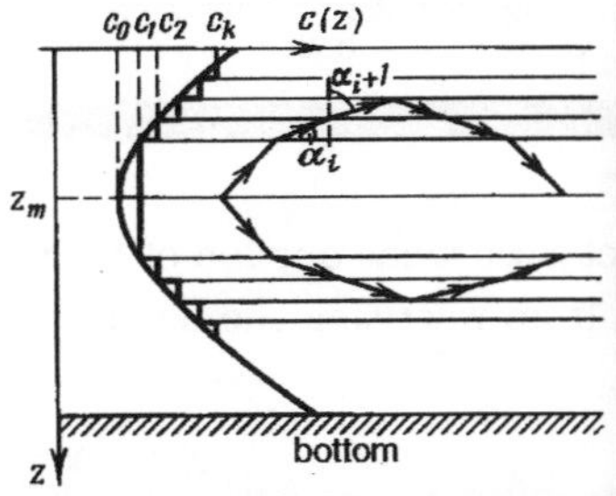

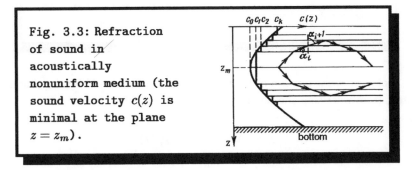

Fig. 3.3: Refraction of sound in acoustically nonuniform medium (the sound velocity $c(z)$ is minimal at the plane $z = z_m$).

Thus, the refraction of sound in the ocean allows a portion of the sonic energy emitted by the source to propagate through the water without rising up to the surface or dropping down to the ocean floor. This means that we have an oceanic acoustic waveguide. The role of "walls" waveguide is played by the layers of water at the depths where the sound beams reflect.

The level z_m where the speed of sound reaches the minimum is called the axis of the waveguide. Usually the depths z_m range from 1.000 to 1.200 meters, but in low latitudes, where the water gets warmed deeper down the axis can descend down to 2.000 m. On the other hand, in high latitudes the influence of temperature on the distribution of the sound speed is noticeable only close to the surface, and therefore the axis rises to the depth of $200 - 500\,m$. In the polar latitudes it lies still closer to the surface.

There are two different types of waveguides in the ocean. The first type occurs when the speed of sound near the surface (c_0) is less than that at the ocean floor (c_f). This usually occurs in deep waters, where the pressure at the bottom reaches hundreds of atmospheres. As we mentioned above, sound reflects well from the water-air interface. So if the ocean surface is smooth (dead calm), it can serve as the upper boundary of a waveguide. The channel then spreads through the entire layer of water, from the surface to the floor (see Fig. 3.4).

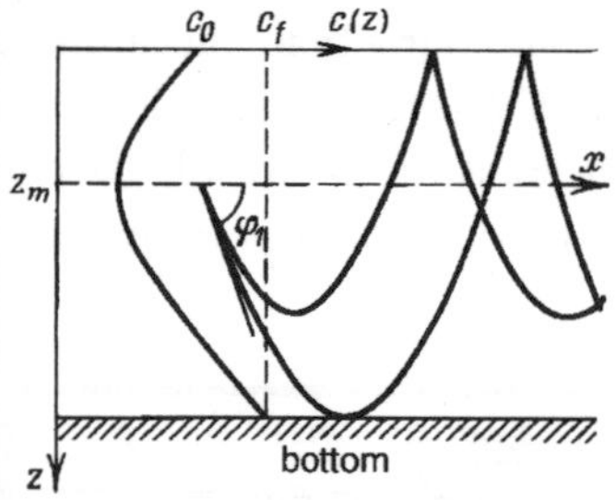

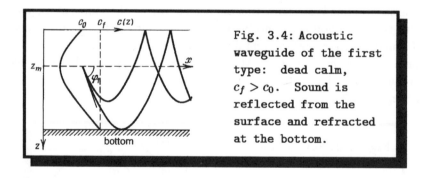

Fig. 3.4: Acoustic waveguide of the first type: dead calm, $c_f > c_0$. Sound is reflected from the surface and refracted at the bottom.

Let's see what portion of the sound beam is "captured" by the channel. Start from rewriting the Snell's law as

$$c(z)\,\cos\varphi_1 = c_1\,\cos\varphi(z).$$

where φ_1 and $\varphi(z)$ are the angles formed by sound rays with the horizontal at depths z_1 and z, respectively. It's clear that $\varphi_1 = \frac{\pi}{2} - \alpha_1$, $\varphi(z) = \frac{\pi}{2} - \alpha(z)$. If the source of sound is located on the axis of the channel ($c_1 = c_m$) then the last sound ray captured by the channel is the one tangent to the ocean floor, $\varphi(z) = 0$, as shown in Fig. 3.4. Therefore all rays that leave the source at angles satisfying the condition

$$\cos\varphi_1 \geq \frac{c_m}{c_f},$$

enter the channel.

When the water surface is rough, all sound beams will scatter from it. The rays leaving the surface at angles larger than φ_1 will reach the floor and be absorbed there. Yet even in this case thanks to refraction the channel can capture those rays that somewhat do not reach the rough surface (Fig. 3.5). Then the channel spreads from the surface to a depth z which can be determined from the condition $c(z_k) = c_0$. It's clear that such a channel captures all sound rays with angles

$$\varphi_1 \leq \arccos\frac{c_m}{c_0}.$$

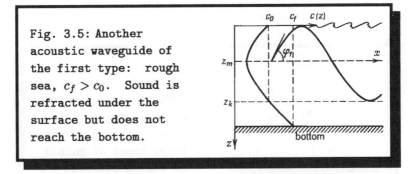

Fig. 3.5: Another acoustic waveguide of the first type: rough sea, $c_f > c_0$. Sound is refracted under the surface but does not reach the bottom.

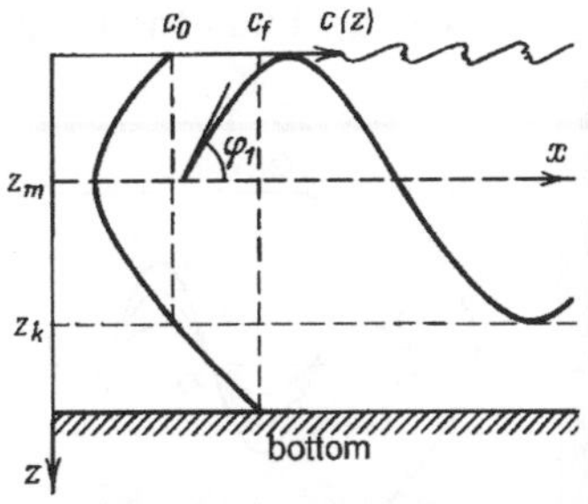

The second type of waveguide is characteristic of shallow water. It occurs only when the speed of sound near the surface is greater than that at the floor, see Fig. 3.6. The channel occupies the water layer from the ocean floor up to the depth z_k where $c(z_k) = c_f$. It looks like a waveguide of the first type flipped upside down.

For certain types of dependencies of sound speed on depth, the waveguide focuses sound beams like a lens. If a sound source is located on the axis, the rays that have leaving it at different angles will periodically converge at some points on the axis. These points are called focuses of the channel. So if the dependence of the speed of sound in the channel on depth is close to parabolic: $c(z) = c_m\,(1 + \frac{1}{2}b^2\,\Delta z^2)$, where $\Delta z = z - z_m$.

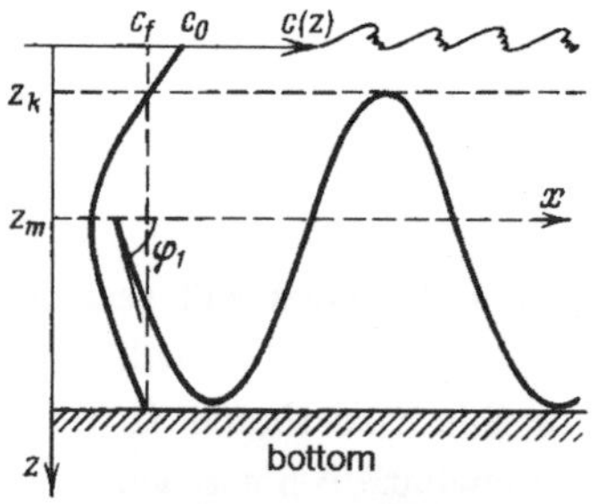

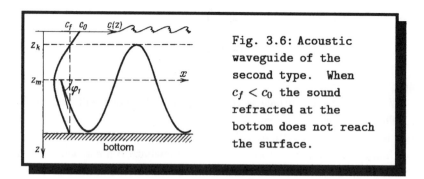

Fig. 3.6: Acoustic waveguide of the second type. When $c_f < c_0$ the sound refracted at the bottom does not reach the surface.

Then for rays leaving the source at small angles with the horizontal, the focuses will lie at the points $x_n = x_0 + \pi n / b$, where $n = 1, 2, \ldots$ and b is a coefficient whose dimension is inverse to depth (m^{-1}), Fig. 3.7. The parabolic function $c(z)$ is close to the actual dependence of the speed of sound on depth in deep oceanic acoustic waveguides. Deviations of $c(z)$ from parabolic law blurs the focuses on the axis of the waveguide[b].

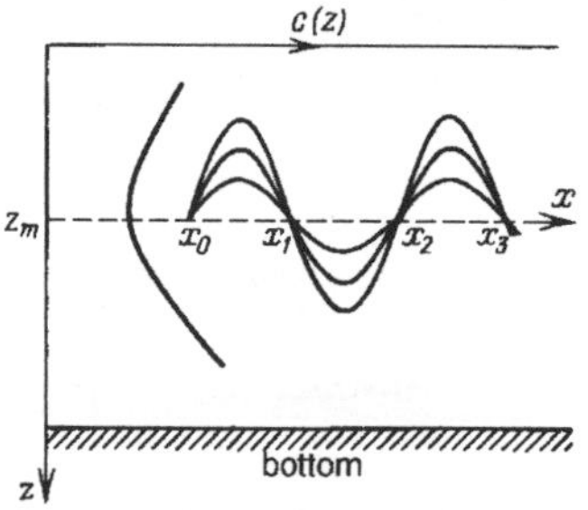

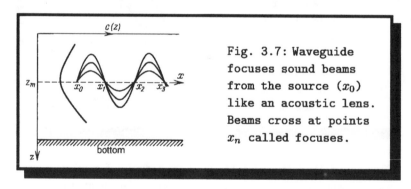

Fig. 3.7: Waveguide focuses sound beams from the source (x_0) like an acoustic lens. Beams cross at points x_n called focuses.

[b]Like many periodic processes in nature propagation of beams along a parabolic waveguide obeys the harmonic law. Near the axis trajectories follow the equation:

$$\frac{d^2\,\Delta z}{dx^2} = -b^2\,\Delta z,$$

where x is not the time but the horizontal coordinate. Obviously the trajectories are sinusoidal, $\Delta z = A \sin b\,(x - x_0)$, cross at the zeros of the sine, $x_n - x_0 = \pi n / b$. —*A. A.*

3.4 Applications?

Is it possible to send a sound signal along an oceanic acoustic waveguide and receive it at the point of origin, after it has completely circled the globe? The answer is a flat no. First and foremost, the continents present insurmountable obstacles, as well as great contrasts in depths of the World oceans. So it's impossible to choose a direction along which there would be a continuous round-the-world waveguide. But that isn't the only reason. A sound wave propagating along an oceanic acoustic waveguide differs from sound waves in the "telephone" tubes on ships that we mentioned at the outset. The sound wave traveling from the bridge to the engine room is one-dimensional, and the area of its wave front is constant at any distance from the source. Therefore, the strength of the sound will also be constant everywhere along the tube (heat losses aren't taken into account). As for the oceanic acoustic waveguide, the sound wave doesn't propagate along a straight line but in all directions in the plane $z = z_m$. So the wave here is a cylindrical surface. Because of this, the strength of the sound decreases with distance — that is, the sound intensity is proportional to $1/R$, where R is the distance from the source of the sound to the detector. (Try to obtain this dependence and compare it with the law of attenuation of a spherical sound wave in three-dimensional space.)

Another reason of sound attenuation is the damping of sound wave as it travels through waters of the ocean. Energy from the wave is transformed into heat due to viscosity of water and other irreversible processes. Moreover, sound waves dissipate in the ocean because of various heterogeneities, such as suspended particles, air bubbles, plankton, and even the swim-bladders of fish.

Finally we should point out that the underwater sound channel isn't the only example of waveguides in nature. Long-distance broadcasting from radio stations is possible only because of the propagation of radio waves through the atmosphere along giant waveguides. And we're sure you've heard of mirages, even if you've never seen one. Under certain atmospheric conditions, waveguide channels for electromagnetic waves in the visible range can form. This explains the a appearance of a ship in the middle of desert, or a city that springs to life in the middle of the ocean.

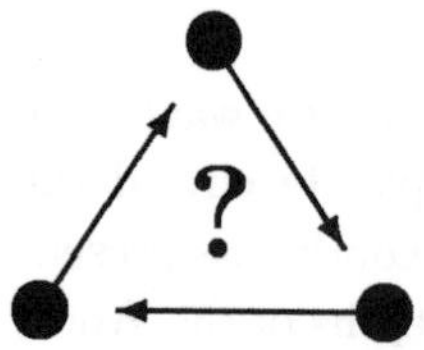

Prove that trajectories of sound beams in parabolic waveguide obey the harmonic equation.

Chapter 4
In the blue

Now the wild white horses play,
Champ and chafe and toss in the spray.

Matthew Arnold, The forsaken merman.

Artists are endowed with sharp professional vision. This makes world in realistic landscapes outstandingly bright and colorful and features some natural events. Even best painters do not need to understand the reasons of the portrayed processes which often may be hidden deeply. However with the help of landscapes of a good painter one can study the ambient nature even better than in physical world. Suspending the moment in the picture the author intuitively focuses on the principals and omits inessential and passing details.

Look at the picture *"In the blue"* by the russian painter Arkadii Rylov[a] that is reproduced on the second page of the cover. "White birds soar among white clouds bathing in the blue sky. And the ship under the sails drifts peacefully over the gently rolling blue waves like a white bird in the ocean." This description belongs to the notable Russian art critic A. A. Fedorov-Davydov. Watching the wonderful canvas in Moscow Tretyakov Gallery one forgets of being in museum and feels a guest at this feast of nature.

But let us leave the aesthetic side and examine the picture with the eyes of researchers. First of all, where from did the author paint the landscape, was he on a rocky ledge on the coast or aboard a ship?

Most probably he was on a ship since there is no surf in the foreground,

[a]A. A. Rylov, (1870–1939), Russian Soviet painter of epic romantic manner.

the distribution of waves is symmetrical and not deformed by the neighboring shore.

Let us try to estimate the velocity of the wind filling the sails of the ship floating at a distance. We are not the first interested by the question of evaluation of the speed of wind on the basis of the height of waves and other natural evidence. It was already in 1806 that Sir Francis Beaufort[b] introduced his approximate twelve-step scale. He related the force of the wind to the effects it had onto land objects and to the sea choppiness, see Table 4.1 on pp. 34–35. This scale has been approved by the International Meteorological Committee and is used up till now.

Turning back to the picture we see that the sea is rather quiet but to rare white horses. According to the Beaufort scale this corresponds to a gentle breeze with the velocity about 10 *mph*.

One may judge velocity of wind not only by means of the Beaufort scale but from the brightness contrast between the sea and the sky. Usually horizon in the open sea looks like a clear-cut border. The brightness of the sea becomes equal to that of the sky only at dead calm. In this case the contrast disappears and the sea becomes indiscernible from the sky. This happens rarely because the calm must be almost absolute, Beaufort number being zero. Even slightest wind disturbs the sea surface. The coefficient of reflection from the oblique elements of the surface is no longer equal to unity and the contrast appears between the sea and sky. It may be studied experimentally. The dependence of the sea-sky contrast on the wind force was measured during an expedition of the russian research ship *"Dmitri Mendeleyev"*. The results of measurements are depicted in Fig. 4.1 by crosses, while the solid line represents the theoretical relation found by A. V. Byalko and V. N. Pelevin.

By the way, why are white watercaps so much unlike the blue-green sea water?

The color of the sea is defined by many factors. Among the most important are the position of the sun, the color of the sky, the form of the sea surface and the depth. In shallow waters presence of sea-weeds and pollution by solid particles are relevant. All these factors affect reflection of light from the surface, it's submarine scattering and absorption. This makes unambiguous prediction of the sea color impossible. Nevertheless

[b]F. Beaufort, (1774–1857), English admiral, hydrographer and cartographer, head of english hydrographic service.

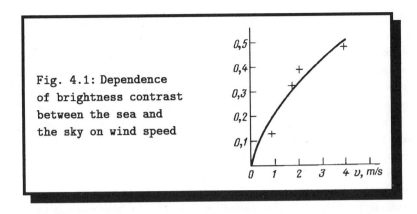

Fig. 4.1: Dependence of brightness contrast between the sea and the sky on wind speed

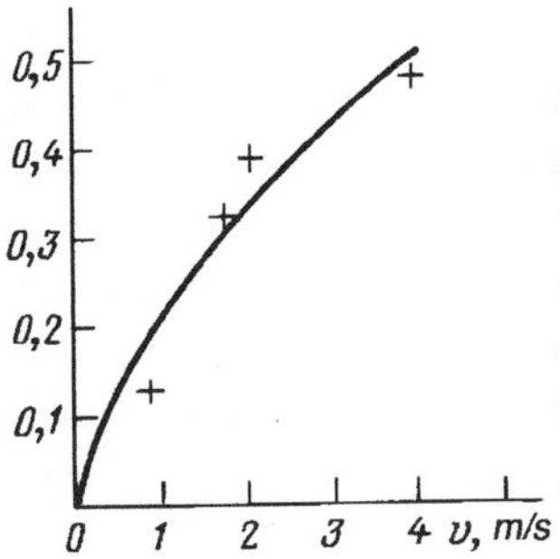

some details can be understood. For example, we can explain why the foreground waves that were closer to the painter were much darker in hue than most of and why the sea became lighter near horizon.

The extent of reflection of a light wave from the interface of two media with different optical densities is determined by the angle of incidence and the relative refraction index of the media. It is quantitatively characterized by the coefficient of reflection that is equal to the ratio of intensities of the incident and reflected rays[c]. Coefficient of reflection depends on incidence. In order to check this you may watch reflection of light from a polished table. The transparent varnish will act as the more dense optical medium. You will see that tangent rays are almost completely reflected, but as the angle of incidence becomes smaller the more light penetrates the optically dense medium and less is reflected from the boundary. Coefficient of reflection falls down with the decrease of the incidence angle.

Let us consider the schematic image of a wave in Figure 1 (see the second page of the cover). It is obvious that the incidence angles α_1 and α_2 of the rays coming to observer from the front and back sides of a wave are different, $\alpha_2 > \alpha_1$. Hence more light comes to the eyes of observer from the distant areas and front faces of waves look darker than the plane sea far away. Certainly in the troubled sea the angle α varies. However far enough the angular size of the darker crests decreases rapidly even though the angle α_2 remains bigger than α_1. Near horizon the observer sees not solitary waves but the averaged pattern, the troughs between waves are

[c]Intensity of light is the averaged over time value of the light flux through a unit area perpendicular to the direction of the light propagation.

hidden and gradually the darker sides of waves disappear. Because of that in the picture the sea near horizon looks lighter than in the foreground.

Now we can explain why watercaps are white. The seething water in the cap swarms with endlessly moving, deforming and bursting bubbles. Reflection angles vary from point to point and with time. As a result sun rays are almost fully reflected by the froth and watercaps appear white[d].

The color of the sea is greatly affected by the color of the sky. We have already said that the first is practically unpredictable. However the second can be understood from physical principles. Obviously the color of the sky is determined by scattering of solar rays in the atmosphere of the Earth. The spectrum of the sunlight is continuous and contains all wavelengths. Why does the scattering make the sky blue whereas the Sun seems yellow? We shall explain that with the help of the Rayleigh law of the light scattering.

In 1898 the English physicist Lord Rayleigh[e] developed the theory of scattering of light by particles much smaller than the wavelength of the light. He found the law that states that ***the intensity of the scattered light varies inversely with the fourth power of the wavelength.*** In order to explain the color of the sky Rayleigh applied his law to the scattering of sunlight in the atmosphere. (For this reason the statement above is sometimes called the "law of the blue sky".)

Let us try to understand the meaning of the Rayleigh law qualitatively. Remember that light consists of electromagnetic waves. Molecules are built of charged particles, *i. e.* of electrons and nuclei. In the field of an electromagnetic wave these particles start moving and the motion may be considered as harmonical: $x(t) = A_0 \sin \omega t$, where A_0 is the amplitude of oscillations and ω is the wave frequency. The acceleration of the particles is $a = -\omega^2 A_0 \sin \omega t$. Accelerated charged particles become sources of electromagnetic radiation themselves and emit the so-called secondary waves. The amplitude of secondary wave is proportional to the acceleration of the emitting particle. (As you know uniformly moving charged particles make electric current but do not generate electromagnetic waves.) Therefore the intensity of the secondary emission is proportional to the square of the ac-

[d]In the froth of watercaps blue rays coming from the sky are mixed with yellowish sunlight giving the white color.

[e]J. W.S. Rayleigh, (1842–1919), English physicist, chairman of the London Royal Society, Nobel Prize 1904.

celeration of electrons in the field of the primary wave (one may neglect the motion of heavy nuclei) and therefore to the fourth power of the frequency, $(I \propto a^2 \propto (x_t'')^2 \propto \omega^4)$.

Now return to the sky. The ratio of the wavelength of blue to that of red is $650\,nm/450\,nm = 1.44$ ($1\,nm$ (nanometer) $= 10^{-9}\,m$). The fourth power of this number is 4.3. Thus according to the Rayleigh law the intensity of the blue light scattered by the atmosphere exceeds that of the red four times. As a result the ten miles thick air layer acquires the blue tint. On the contrary the blue component of the sunlight that reaches us through the atmospheric "filter" is seriously depleted. Hence the sunrays penetrating the atmosphere get the yellowish tone. The coloration may enhance getting red and orange during sunset when the rays have to pass longer path through the air. (Obviously the colors change in the reverse order when the sun rises.)

It is interesting that despite the Rayleigh law claims the wavelength of the scattered light to be much bigger than scattering particles the intensity of scattering does not depend on the particle size. At first Rayleigh believed that the color of the sky is due to the tiniest dust polluting the atmosphere. But then he decided that sunrays are scattered by molecules of gases that make up the air. Ten years later in 1908 the Polish theoretical physicist M. Smoluchowski[f] proposed the idea that scattering is effected by rather unexpected objects, namely by inhomogeneities of the density of particles. With the help of this hypothesis Smoluchowsky managed to explain the long known phenomenon of ***critical opalescence*** — that is the strong scattering of light in liquids and gases that occurs near critical point. Finally in 1910 Albert Einstein[g] formulated the consistent quantitative theory of molecular scattering of light that leaned upon the ideas of Smoluchowski. In case of gases the intensity of scattered light exactly coincides with the earlier result by Rayleigh.

It seems that everything is in order now. But what is the origin of inhomogeneities in the air? Supposedly air must be in thermodynamical equilibrium. Gigantic inhomogeneities that make wind blow are incomparable with wavelengths of light and can not affect the scattering.

In order to clear out the origin of the inhomogeneities we must discuss the concept of thermodynamical equilibrium in more detail. For simplicity

[f]M. von Smolan-Smoluchowski, (1872-1917), Polish physicist, classical works on fluctuation theory, and theory of Brownian motion

[g]See footnote on page 5

let us consider some macroscopic amount of gas confined in a closed volume.

Physical systems consist of enormous numbers of particles. This makes statistical description the only possible approach. Statistical treatment means that instead of following separate molecules we calculate average values of physical characteristics of the whole system. It is not necessary at all that the corresponding value was the same for all molecules. The most probable distribution of molecules in the macroscopic gas volume would be the uniform one. But because of thermal motion there is a nonzero probability that concentration of molecules in some part of our container will enhance (and as a result it will fall down somewhere else). Theoretically it is even possible that all molecules will assemble in one half of the container leaving another half absolutely empty. However the probability of such event is expressed by the extremely small number. So there is little hope to realize it once in 10^{10} years that is currently believed to be the lifetime of the Universe.

But small deviations of physical quantities from their averages are not only allowed but happen perpetually due to the thermal motion of molecules. These deviations are called ***fluctuations;*** (in latin ***fluctus*** means wave). Because of fluctuations the gas density may be greater here and less there and as a result the refraction index will vary from place to place.

If now we turn back to the scattering of light. All the reasoning applied to the closed container will hold in the atmosphere. Light is scattered by inhomogeneities of the refraction index that come from density fluctuations. Moreover, air is the mixture of several gases. Distinctions in thermal motion of different molecules provide one more source of inhomogeneities.

The typical scale of inhomogeneities of the refraction index (and of those of density) depends on temperature. The sunlight is mainly scattered in the atmospheric layers where inhomogeneities are much less than wavelengths of visible light but much greater than the molecules of gases that compose the air. This means that scattering is effectuated by inhomogeneities but not by molecules as it was presumed by Rayleigh.

Nevertheless the sky is blue and not violet contrary the prediction of the Rayleigh law. There are two reasons for this discrepancy. First, the spectrum of the Sun contains much less violet rays than blue ones. The second thing responsible for the seeming disagreement of the theory with practice is our "registrating device", the ordinary human eye. It turns out that visual perception markedly depends on the wavelength of light. The experimental curve characterizing this dependence is plotted in Fig. 2 of the second page of the cover. It is clear that our eye responds to violet rays much weaker than to blue and green ones. This conceals the violet component of the scattered sunrays from people.

Well, but why are the clouds that we see in the sky unmistakably white? Maybe they are composed of particles that violate the Rayleigh law and our conclusions are no longer true?

Clouds consist either of water drops or of ice crystals that in spite of being microscopic are much larger than the visible wavelengths. The Rayleigh law is not valid for these particles and the intensity of the scattered light is nearly the same for all wavelengths. This makes clouds like those painted in the picture.

Now turn your attention to the form of the clouds. The tops of the clouds in the picture are drawn fluffy and wreathing (these are the so-called heap clouds) but the lower surfaces are plainly outlined. For what reason? It is known that heap clouds (in distinction to sheet ones) are formed by uprising convective flows of humid air. The temperature of air decreases with height above the sea (as well as above the land). As long as the altitude is much less than the Earth radius and the distance to the nearest shore surfaces of constant temperature (the isotherms) are next to horizontal plains parallel to the sea surface. The temperature drop is sufficiently fast near the sea level, being about 1°C per hundred meters (this makes about = 1.6°F per 100 yd). (Generally speaking the dependence of the air temperature on the altitude is far from linear but for elevations less than several miles these numbers are correct).

Now, what happens to the upward flow of humid air? As soon as it reaches the altitude where the temperature of air corresponds to the dew point of the vapor it carries water begins condensing to tiniest drops. The isotherm where this happens contours the bottom of the cloud. Irregularities of the surface where condensation takes place stay within some tens of meters whereas clouds stretch to hundreds and thousands meters. Hence their lower bounds are almost flat. This is confirmed by the row of clouds far away hovering horizontally above the see.

However the rising does not stop with creating the bottom of a cloud. The air continues the upward motion and cools rapidly. The remaining vapor suffers intensive condensation first into droplets and then into little ice crystals. These crystals usually form the top of heap clouds. After having lost the vapors and cooling the air slows down and turns back. It flows sidewards and around the cloud. Convective flows lead to formation of the characteristic curls at the tops of heap clouds and the descending cold air keeps clouds apart. Thanks to that they do not merge into heavy grey mass being interspaced by blue intervals.

You see in the foreground the flight of white birds. Let us estimate the frequency of wing flaps of a medium-sized bird (say with the mass $m \approx 10\,kg$ and the area of wings $S \approx 1\,m^2$) when it flies without gliding. Let the average velocity of the wings be $\bar{v}$. Then in the time Δt the bird wings will render velocity $\bar{v}$ to the mass $\Delta m = \rho\, S\, \bar{v}\, \Delta t$ of the air (here ρ is the air density). The total momentum passed by the wings to the air will be $\Delta \bar{p} = \rho\, S\, \bar{v}^2\, \Delta t$. In order to keep the height this must compensate the weight of the bird. This means that $\Delta \bar{p} = m\, g\, \Delta t$. We may conclude that

$$m\, g = \rho\, S\, \bar{v}^2,$$

and the mean velocity of the moving wings is $\bar{v} = \sqrt{m\, g / \rho\, S}$. This velocity can be related to the frequency ν of flaps and the length of the wings L in the usual way:

$$\bar{v} = \omega\, L = 2\pi\, \nu\, L.$$

Assuming that $L \sim \sqrt{S}$ we find:

$$\nu \approx \frac{1}{2\pi\, S} \sqrt{\frac{m\, g}{\rho}} \approx 1\, s^{-1}. \tag{4.1}$$

Thus according to our calculation the bird should flap the wings once per second that looks quite a reasonable estimate by the order of magnitude.

It is interesting to discuss the obtained formula in more detail. Let us suppose that all birds have roughly the same form of the body regardless to the size and species. Then one may link the area of the wings to the mass of the bird by the relation $S \propto m^{2/3}$. Substituting this into the earlier found expression for the frequency of the wing flaps we obtain that

$$\nu \propto \frac{1}{m^{1/6}}.$$

From this we conclude that the frequency of flaps grows with the decrease in the mass of birds. This absolutely agrees with the common sense. Certainly the assumption that all birds have wings of the same form is extremely rough since wings of most of big birds are relatively bigger than those of small ones. Nevertheless this only supports the trend.

Let us note that the same formula (4.1) could be derived using the dimensional approach (except the important factor 2π). It is clear that the frequency of wing flaps depends on the bird's weight, the area of the wings S, and the density ρ of the ambient air. Let us search for a relation between the four of them. Suppose that $\nu = \rho^{\alpha} S^{\beta} (m g)^{\gamma}$ with α, β, γ being unknown numbers. Comparison of the dimensions of the quantities in the both sides of the relation gives $\alpha = -\gamma = -1/2$ and $\beta = -1$. From here follows that

$$\nu \sim \frac{1}{S}\sqrt{\frac{m g}{\rho}}.$$

Questions and answers that could be found ***"in the blue"*** are far from being exhausted. Curious and observant reader will find in the picture other maybe even more instructive aspects. However why should we limit ourselves by the frame of the picture? There is plenty of interesting questions and problems in the everyday world all around us.

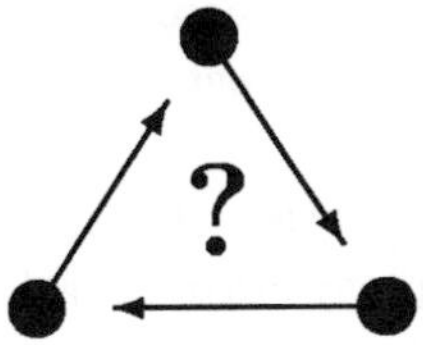

By the way could you now explain why astronauts tell that from the outer space the Earth looks blue?

Table 4.1: BEAUFORT WIND SCALE.

Beaufort Force Number	State of Air	Wind velocity		Description of sea surface
		(knots)	(mph)	
0	calm	0-1	< 1.15	sea like a mirror
1	light airs	1–3	1.2–3.5	ripples with appearance of scales are formed, without foam crests
2	slight breeze	4–6	4.6–6.9	small wavelets still short but more pronounced; crests have a glassy appearance but do not break
3	gentle breeze	7–10	8.0–12	large wavelets; crests begin to break; foam of glassy appearance; perhaps scattered white horses
4	moderate breeze	11–16	13–18	small waves becoming longer; fairly frequent white horses
5	fresh breeze	17–21	20–24	moderate waves taking a more pronounced long form; many white horses are formed; chance of some spray
6	strong breeze	22–27	25–31	large waves begin to form; the white foam crests are more extensive everywhere; probably some spray
7	moderate gale	28–33	32–38	sea heaps up and white foam from breaking waves begins to be blown in streaks along the direction of the wind; spindrift begins to be seen
8	fresh gale	34–40	39–46	moderately high waves of greater length; edges of crests break into spindrift; foam is blown in well-marked streaks along the direction of the wind

Table 4.1: BEAUFORT WIND SCALE *(continued)*.

Beaufort Force Number	State of Air	Wind velocity		Description of sea surface
		(knots)	(mph)	
9	strong gale	41–47	47–54	high waves; dense streaks of foam along the direction of the wind; sea begins to roll; spray affects visibility
10	whole gale	48–55	55–63	very high waves with long overhanging crests; resulting foam in great patches is blown in dense white streaks along the direction of the wind; on the whole the surface of the sea takes on a white appearance; rolling of the sea becomes heavy; visibility affected
11	storm	56–65	64–75	exceptionally high waves; small- and medium- sized ships might be for a long time lost to view behind the waves; sea is covered with long white patches of foam; everywhere the edges of the wave crests are blown into foam; visibility affected
> 12	hurricane	> 65	> 75	the air is filled with foam and spray; sea is completely white with driving spray; visibility very seriously affected

Chapter 5

The moon-glades

Reflections of various light sources from the surface of water often look like long shimmering lanes leading from the source to our eye. Just remember the setting sun reflected by sea or street lights along a night river quay. Glittering of the moon adorns the sea or lake with a wide stripe of light.

All this happens because every wavelet on the surface gives a separate image of the source. Let us try to understand why reflections from thousands of illuminated ripples make a glade, that is an oblong figure directed from the light source to the observer.

As you already know wavelets are formed at Beaufort numbers between 1 and 3. At weaker winds water is calm and the surface reflects like a plain mirror. Stronger winds bring on foam and white horses and the contour of the glade becomes vague. One may visualize ripples as scores of wavelets running chaotically in all directions[a]. Slopes of their surfaces do not exceed some limiting value α which depends on the wind and can reach 20°–30°.

Now let us a bit simplify the problem and try to find proportions of the glade. Suppose that everywhere on the surface there is plenty of tiny mirror-like ripples looking in all directions. The slopes of wavelets range from 0 to α (this follows from the smoothness of the surface). For simplicity we shall assume that both the light source and the observer are at the same height h above the water, Fig. 5.1.

A small horizontal mirror will cast the beam to the eye of the observer only if its distances to the source and to the observer are equal. This means it must be at the point M. On the other hand a mirror tilted at the angle

[a]Remember the Raileigh independence principle —*A. A.*

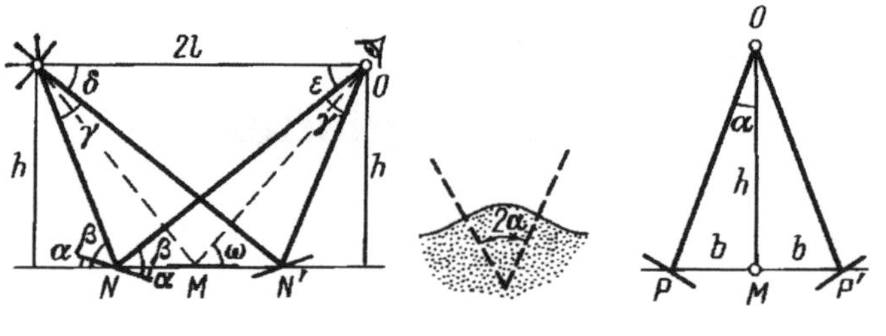

Fig. 5.1: The length of glade is determined by the steepest slopes turned to and away from the observer. The width of glade is defined by reflections from slopes tilted sidewards.

α towards the observer must be moved away from him to the point N. Conversely a mirror tilted away from the observer should be transferred to the closer point N'.

The tilted mirrors imitate the extreme positions from where waves still reflect light to our eyes. Therefore the distance from N to N' determines the length of the light-glade. Everywhere in between one can find waves with the right slope which reflect rays to the observer.

Now let us consider angles between the light rays. One can notice from the Fig. 5.1 that $\beta+\alpha=\gamma+\delta$, $\beta-\alpha=\varepsilon=\delta$ and $\gamma=\alpha+\beta-(\beta-\alpha)=2\alpha$. Hence we may conclude that the angular size of the longer axis of the light spot simply equals the angle between the steepest slopes of the ripples. It makes no problem to calculate the length $N\,N'$ of the glade.

The shorter axis of the reflecting light spot is easily calculated in the similar manner. Let us shift the mirror away from the central point M in the direction transverse to $N\,N'$. In order to reflect beams to the eye the mirror must be turned around the axis parallel to $N\,N'$, Fig. 5.1. Having in mind that the maximal tilt of the mirror remains α we find that the width of the light strip is $P\,P' = 2h\,\tan\alpha$ and hence the shorter axis is seen at the angle $\beta = 2\arctan\frac{h\,\tan\alpha}{\sqrt{l^2+h^2}} \approx \frac{2h\,\tan\alpha}{\sqrt{l^2+h^2}}$.

The ratio of the apparent sizes of the two axes is $\beta\,/\,2\alpha$. In case that the spot is not too large and the angle α is small then $\beta\,/\,2\alpha \approx \frac{h}{\sqrt{l^2+h^2}} = \sin\omega$

where ω is the incidence angle at the point M.[b] As the angle ω becomes smaller the spot grows longer. If the look glances along the surface the spot appears infinitely long and narrow.

When we watch moon-glades on sea the angle ω is most often small and the glade stretches to the horizon, Fig. 5.2. Of course here our formulae are not literally applicable. Nevertheless they help to explain qualitatively the origin of glades and understand the effects of wind speed and of the moon altitude above horizon on glade width: increasing α and h makes the glade wider.

Fig. 5.2: The wind speeds are (from left to right): $12m/s$; $12m/s$; $5m/s$; $2m/s$. The altitudes of the sun above horizon are: $30°$; $20°$; $13°$; $7°$.

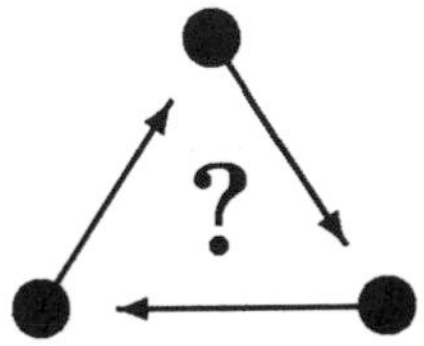

What is the value of the angle ω if the angular altitude of the moon is β?

[b]To be specific M is the point where the reflection from the calm waters would be. —*A. A.*

Chapter 6

The Fucault pendulum and the Baer law

> ... There was something, however, in the appearance of this machine which caused me to regard it more attentively.
>
> *Edgar Allan Poe,* The Pit and the Pendulum.

Those lucky ones who have been to St. Petersburg must remember the famous pendulum in the St.Isaac's cathedral[a]. Others might have heard about it, fig. 6.1. The swings of the pendulum are accompanied by the slow rotation of the plane of oscillations. This observation was first done in 1851 by the French scientist J. Fucault[b]. The experiment was carried out in the spacious hall of the Pantheon in Paris, the ball of the pendulum had the mass of 28 *kg*, (68 *lb*) and the string was 67 meters (73 *yd*) long. Since then this sort of pendula are called after Fucault. How could one explain it's motion?

You know from the textbooks that if the Newton's[c] laws were true on the Earth then the pendulum would keep the plane of oscillations. This means that in the reference frame rigidly bound with the Earth the laws of Newton must be "corrected". In order to do this one has to introduce

[a]St. Petersburg, the city and seaport on the Gulf of Finland of the Baltic Sea was during 1712-1917 the capital of Russian Empire. In 1917, after revolution, it was renamed to Leningrad but now bears the original name.

[b]J. B. L. Fucault, (1819–1868), French physicist, foreign member of Russian Academy of Sciences.

[c]Sir Isaac Newton, (1642–1727), English philosopher and mathematician; formulator of the laws of classical mechanics and gravity.

Fig. 6.1: In March 1931 Fucault pendulum was first presented in the Isaac's cathedral occupied at that time by Leningrad Antireligious Museum.

special forces called ***inertia forces.***

6.1 Inertia forces in the rotating reference frame

Inertia forces must be introduced in any reference frame that moves with acceleration with respect to the Sun (or, to be precise, to the so-called stationary stars). These are called non-inertial reference frames in distinction to inertial ones that move uniformly with respect to the Sun and stationary stars.

Strictly speaking the Earth does not present an inertial frame of reference because it orbits the Sun and revolves itself. Usually one may neglect accelerations arising from these motions and apply the laws of Newton. However Fucault pendulum is not the case. Precession of the oscillation plane is explained by the action of the special force of inertia called the Coriolis force[d]. Let us dwell on it.

Here is the simple example of rotating reference frame where inertia forces reveal themselves clearly.

Imagine a man riding a merry-go-round in the Gorky Park[e]. Let the radius of the circle be r and the angular velocity of rotation ω. Suppose that the man tries to jump from his seat to the one in front of him, Fig. 6.2, moving with velocity v_0 with respect to the platform.

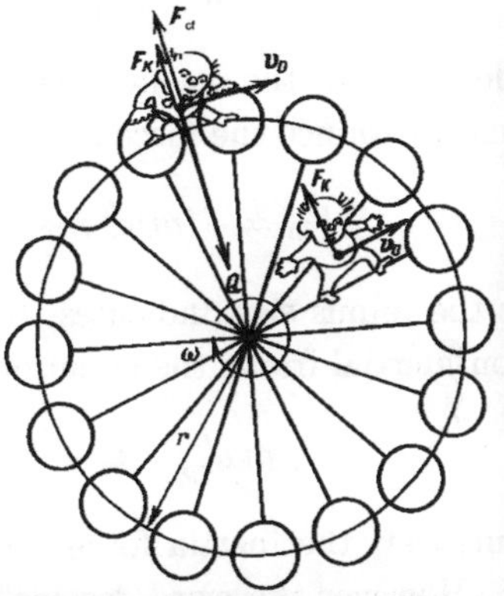

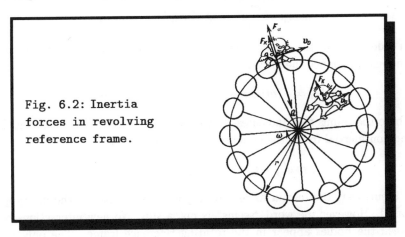

Fig. 6.2: Inertia forces in revolving reference frame.

Warning! The experiment is purely imaginary being strictly forbidden by safety regulations.

First let us consider the motion of our hero in a stationary reference frame. Obviously the motion is circular with the linear velocity v that adds up of the linear velocity ωr of the merry-go round and his relative velocity:

$$v = \omega r + v_0.$$

[d] G. G. Coriolis, (1792–1843), French civil engineer.

[e] See the novel by M. Cruz-Smith for further reference.

The centripetal acceleration is defined by the common formula,

$$a_{\text{cp}} = \frac{v^2}{r} = \frac{v_0^2}{r} + \omega^2 r + 2\,v_0\,\omega.$$

According to the second Newton law the acceleration is due to the horizontal component of the force exerted onto the man by the rotating platform, the seat, handles *etc.*,

$$m\,a_{\text{cp}} = Q.$$

Now consider the motion in the reference frame bound to the merry-go-round. Here the linear velocity is v_0 and the centripetal acceleration is $a'_{\text{cp}} = \frac{v_0^2}{r}$. With the help of the two previous equalities we may write:

$$m\,a_{\text{cp}} = \frac{m\,v_0^2}{r} = Q - m\,\omega^2\,r - 2m\,v_0\,\omega.$$

In order to apply the second Newton law in the revolving frame of reference we must introduce the force of inertia:

$$F_{\text{in}} = -(m\,\omega^2\,r + 2m\,v_0\,\omega) = -(F_{\text{cf}} + F_{\text{Cor}}),$$

where the minus sign indicates that it is directed away from the axis. In the non-inertial frame the equation of motion will be:

$$m\,a'_{\text{cp}} = Q + F_{\text{in}} = Q - (F_{\text{cf}} + F_{\text{Cor}}).$$

It seems that the inertia force throws you of the center of the merry-go-round. However the word "seems" is not a slip. No new interactions between the bodies appear in the rotating reference frame. The only real forces acting onto the man are the same reactions of the seat and bars. Their net horizontal component Q is directed towards the center. In the stationary reference frame the force Q resulted into the centripetal acceleration a_{cp}. In the rotating frame due to kinematical reasons the acceleration changed to the smaller value a'_{cp}. In order to restore the balance between the two sides of the equation we had to introduce the force of inertia.

In our case the force F_{in} comes up of the two addends. The first is the centrifugal force F_{cf} that increases with the frequency of rotation and with the distance from the center. The second is the Coriolis force F_{Cor} named after the person who first calculated it. This force has to be introduced only when the body moves relative to the rotating frame. It depends not on the position of the body but on it's velocity and the angular velocity of rotation.

If the body in the rotating frame moves not along a circle but radially, Fig. 6.2, then, just the same, one must introduce the Coriolis force. Now it is perpendicular to the radius unlike the previous case. One of the basic features of the Coriolis force is that it is always perpendicular both to the axis of rotation and to the direction of motion. It may look strange but in the revolving frame inertia forces not only push a body away from the center but tend to swerve it astray.

We must emphasize that the Coriolis force like all other inertia forces is of kinematical origin and can not be related to any physical objects[f]. Here is an explicit example.

Imagine a cannon set at the North pole and pointed along a meridian (the pole is chosen for simplicity). Let the target lie on the same meridian. Is it possible that the projectile hits the target? From the point of view of external observer which uses the inertial frame bound to the Sun the answer is obvious: the trajectory of the projectile lies in the initial meridional plane whereas the aim revolves with the Earth. Thus the projectile will never get the target (unless a whole number of days will elapse). But how could one explain the fact in the reference frame bound to the Earth? What causes the projectile stray from the initial vertical plane? In order to restore consistency one has to introduce the Coriolis force that is perpendicular to the rotation axis and to the velocity of a body. This force pulls the projectile away from the meridional plane and it misses the target.

Now let us return to the precession of the oscillation plane of the Fucault pendulum from which we have started. It comes of a quite similar reason. Suppose again that the pendulum is situated at the pole. Then for a stationary observer the oscillation plane is at rest and it is the Earth that rotates. A denizen of the North pole will see the opposite. For him the meridional plane looks fixed whereas the oscillation plane of the pendulum performs a full revolution every 24 hours. The only way to explain this is with the help of the Coriolis force. Unfortunately in general the picture is not so transparent as at the pole[g].

[f]Even though the force of inertia is not produced by any real bodies, observers feel it as a real force, akin to gravity. Remember the centrifugal force in the turning car.

[g]Oscillation plane of a Fucault pendulum located elsewhere turns $2\pi \sin \alpha$ radians per day, where α is the latitude of the place.

6.2 Interesting consequences

The Coriolis force which appears due to the Earth rotation leads to a number of important effects. But before discussing those let us establish the direction of the Coriolis force. We have told that it is always perpendicular to the rotation axis and to the velocity of motion. However this leaves two possibilities depicted in Fig. 6.3. Let us remind that analogous situation emerges when defining the direction of the Lorentz[h] force exerted onto a moving charged by magnetic field. You may remember from textbooks that it is perpendicular to the velocity of the charge and to the magnetic induction. Still in order to define it unambiguously one has to resort to the *left hand rule*.

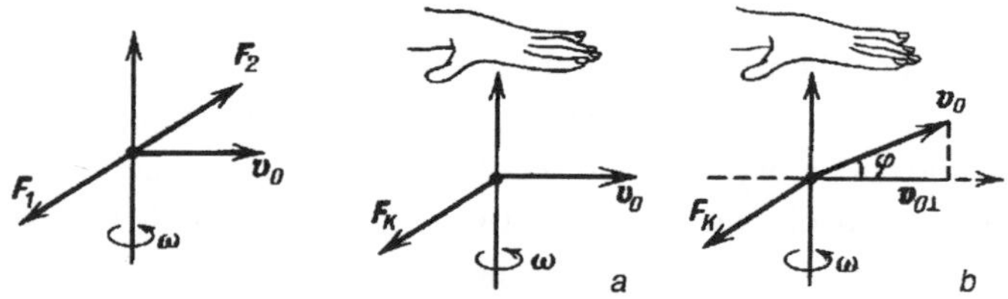

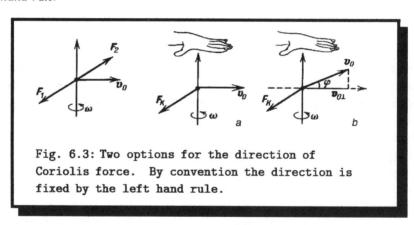

Fig. 6.3: Two options for the direction of Coriolis force. By convention the direction is fixed by the left hand rule.

Direction of the Coriolis force can be determined by means of the similar rule elucidated in Fig. 6.3, *a*. First of all we must assign a direction to the axis. By convention looking in this direction one sees the clockwise rotation[i]. Now let us pose the left hand with the four fingers pointing in the direction of the velocity so that the axis pierced through the palm. Then thumb the set aside at the right angle will show the direction of the Coriolis force.

The alternatives in defining the directions of Coriolis and Lorentz forces correspond to the two kinds of symmetry encountered in nature, the left

[h]H. L. Lorentz, (1853–1928), the Dutch physicist; Nobel Prize 1902.

[i]The *gimlet* rule states that this is where a gimlet rigidly attached to the frame would move.

and right symmetries. In order to classify the symmetry one has every time to use "standards" such as hand, gimlet, cork-screw *etc.* Certainly nature does not care about your hand or gimlet. Simply these are tools that help in fixing the direction of the force.

This completeness the discussion of the Coriolis force for the case when the velocity of a body in the revolving frame is perpendicular to the axis. The magnitude of the force is $2m\,\omega\,v_0$ and the direction is defined by the left hand rule. But what happens in general case?

It turns out that if the velocity v_0 makes an arbitrary angle with the rotation axis, Fig. 6.3, *b*, then only the projection of v_0 onto the plane perpendicular to the axis is important. The value of the Coriolis force is given by the following formula:

$$F_{\text{Cor}} = 2m\,\omega\,v_{\perp} = 2m\,\omega\,v_0 \cos\phi.$$

Direction of the force is determined by the same left hand rule although now the fingers must be parallel not to the velocity but to it's projection onto the plane perpendicular to the axis, Fig. 6.3, *b*.

Now we have learned everything about Coriolis force: both how to calculate the value and to define the direction. Armed with this knowledge we may explicate a number of interesting effects.

Say, it is well known that trade winds which blow from tropics to the equator are always deflected westward. This effect is explained in Fig. 6.4. First let us consider the Northern hemisphere where trade winds blow from north to south. Position the left hand above a globe, the palm down. The axis enters the palm being perpendicular to the fingers. You will that the Coriolis force is perpendicular to the page being levelled at you, that is to the west. Trade winds of the Southern hemisphere in their own turn blow from the tropic of Capricorn north to equator. However neither the direction of rotation nor the projection of wind velocity onto the equatorial plane change. Therefore the direction of the Coriolis force does not change either and in both cases the winds are diverted to the west.

Figure 6.5 illustrates the Baer[j] law. The right banks of rivers in the Northern hemisphere are more steep and undermined than the left ones (and *vice versa* in the Southern hemisphere). The reason is again the Coriolis force that pushes flowing water to the right. Because of friction

[j]K. E. von Baer, (1792–1876), Estonian zoologist and pioneer embryologist was among the founders of the Russian Geographical Society.

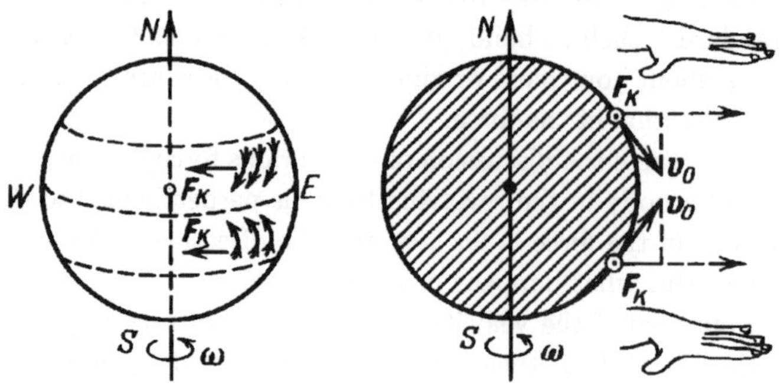

Fig. 6.4: Coriolis force deflects trade winds to the west.

the surface velocity of a stream is bigger than that at the bottom; hence bigger is the Coriolis force. This gives rise to the circulation of water shown in Fig. 6.6 by arrows. The soil of the right bank is washed away and settles at the left side. This strongly resembles wearing of the bank at river turns which was described in the chapter dedicated to meanders.

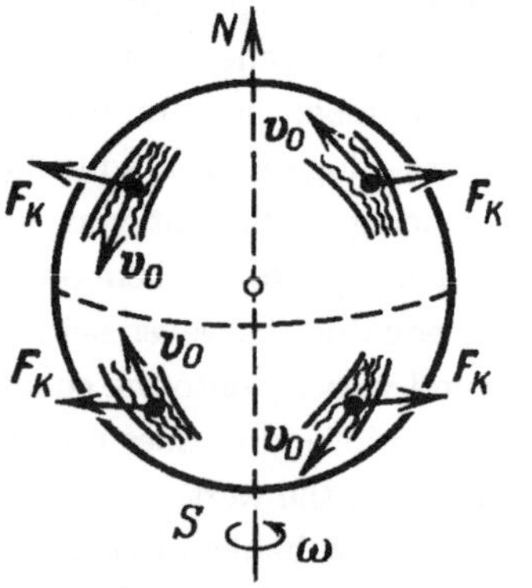

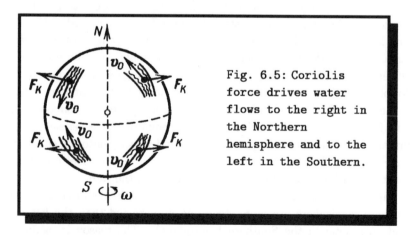

Fig. 6.5: Coriolis force drives water flows to the right in the Northern hemisphere and to the left in the Southern.

Coriolis force leads to eastward deviation of falling bodies. (Tackle this yourself.) In 1833 the german physicist Ferdinand Reich carried out precision experiments in the Freiburg mine. He obtained that the average

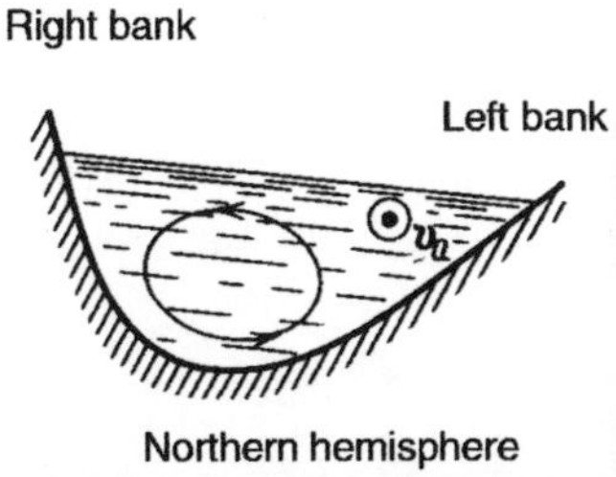

Fig. 6.6: The right banks of rivers in the Northern hemisphere are more steep and undermined than the left ones.

(over 106 measurements) deflection of bodies which were dropped from the height of 158 *m* (178 *yd*) was 28.3 *mm* (1.11 *in*). This was one of the first experimental proofs of the Coriolis theory.

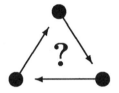

Try to estimate the difference of water levels at the right and left sides of the Volga river.
Does the Baer law apply to rivers streaming along parallels or equator? What changes if a river crosses equator like Congo?

Chapter 7

The moon-brake

Time and tide stop for no man.

A proverb.

It was already long ago that people identified the Moon as the reason of tides. The Moon attracts the water of the world ocean and that forms in the ocean a water "hump". The hump keeps its place on the moon side as the Earth rotates about the axis. When the high water advances to a coast the tide rises and when it retreats the ebb starts. The theory looks rather natural but it leads to a contradiction. This would mean that tides must be a daily event but instead of that they happen every twelve hours.

The first explanation was given by the Newton's theory of tides which appeared short after the discovery of the law of gravity. We shall study this question using the idea of inertia forces. According to the previous chapter one may apply the Newton laws of mechanics in revolving reference frame after adding to interactions between physical bodies forces of inertia.

The Earth rotates around its axis, around the Sun and around... the Moon. Usually on forgets the latter one but it is this rotation that makes possible to construct the theory of tides. Imagine that two balls, one light and one heavy, linked by a string are placed on a smooth surface, Fig. 7.1.

Rotations of the tied balls are interrelated. Each one follows a circle of its own radius but the common center of the two is at the center of mass of the system. Of course the bigger ball traces the smaller circle but it moves! Just the same the Moon and the Earth being attracted to each other according to the law of gravity are orbiting their common spatial

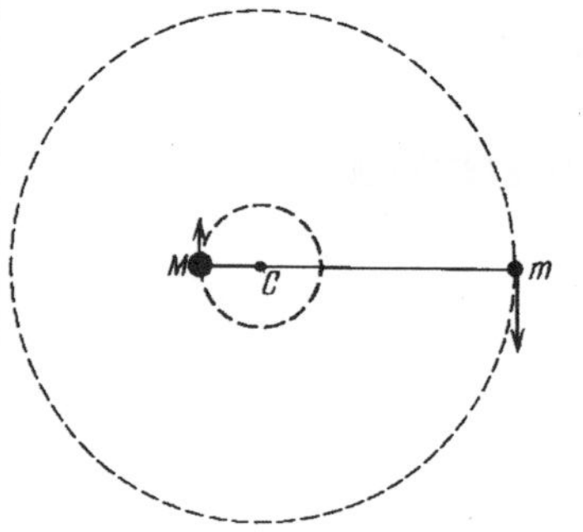

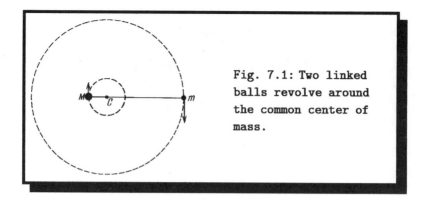

Fig. 7.1: Two linked balls revolve around the common center of mass.

center of inertia C, Fig. 7.2. Because of the big mass of the Earth this point lies inside the globe being shifted with respect to the center O. The angular velocities of rotation around C of both the planet and the satellite are evidently the same.

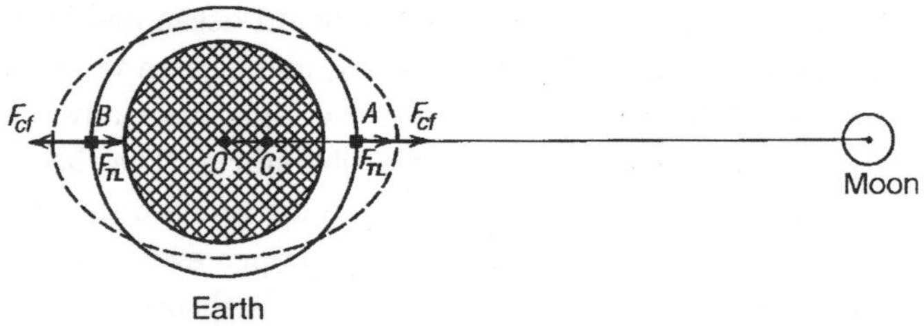

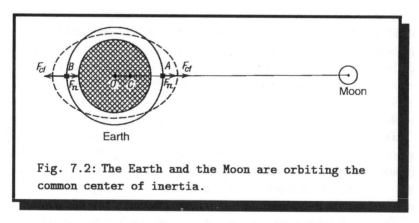

Fig. 7.2: The Earth and the Moon are orbiting the common center of inertia.

Consider now the revolving reference frame where both the Earth and the Moon stay at rest. Inasmuch as the reference frame is noninertial every mass element experiences not only the force of gravity but a centrifugal force as well. The farther from the center C the stronger this force becomes.

Let us imagine for simplicity that water is evenly spread over the entire surface of the globe. May this be an equilibrium? Obviously, not. Gravitational attraction to the Moon and centrifugal forces will destabilize the state. On the moon side the two forces are directed away from the Earth center and give rise to the water hump A, Fig. 7.2. But the situation on the remote side is quite alike. As we step away from the common center of mass the centrifugal force increases whereas the attraction to the Moon falls down. The resultant force is again directed away from the center of the Earth and gives origin to the second hump B. The equilibrium configuration is represented in Fig. 7.2.

Of course this explanation of tides is much simplified. It does not take into account the nonuniform distribution of water on the Earth, effects of attraction to the Sun and many other factors that may essentially influence the picture. Still the theory answers the chief question. Once the humps do not move (with respect to the rotating frame of reference) but the planet revolves around its own axis the tides must occur twice a day.

Now it is the time to explain the principle of the moon-brake. It turns out that the humps actually lie not on the line connecting the centers of the Earth and the Moon (as it was shown for simplicity in Fig. 7.2) but are a little displaced, Fig. 7.3. The reason is that because of friction the ocean rotates together with the Earth. Therefore the mass of water in the humps is continuously renewed. However the deformation is always in retard with respect to the force that brings it on. (The force gives rise to an acceleration but it takes time for particles to gain speed and reach the place.) Thus the top of the hump that is the point of the highest tide is not at the point of the strongest attraction to the Moon which lies on the line connecting the centers. The hump is formed with a lag and it is shifted in the direction of the daily rotation of the Earth. According to the Fig. 7.3 this implies, that the force of gravitational attraction to the Moon does not pass through the center of the Earth and brings about a torque that slows down the gyration. The duration of the revolution daily enhances! This was first recognized by the wonderful English physicist Lord Kelvin[a].

The "moon-brake" operates flawlessly for many millions of years and has the capability to notably change the length of the day. Scientists discovered in corals that have lived in the ocean about 400 millions years ago structures called "diurnal" and "annual" rings. When the diurnal rings were counted it turned out that there were 395 of them per year! The length of a year that is the period of the circumvolution of the Earth around the Sun, probably, did not change since then. Hence at those times the day lasted only 22 hours!

[a]W. Thomson, (1824-1907), 1st Baron Kelvin since 1892; English physicist and mathematician, chairman of the London Royal Society.

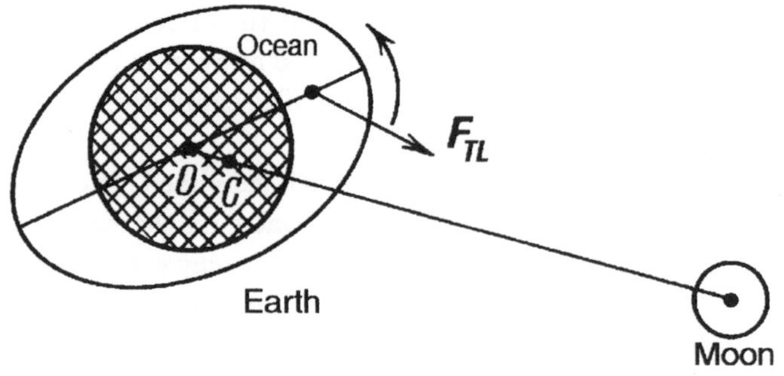

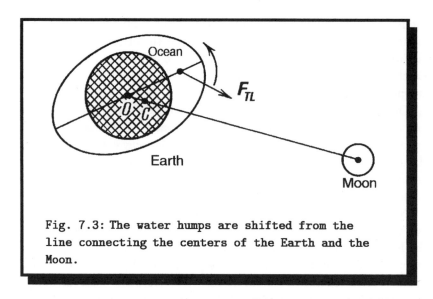

Fig. 7.3: The water humps are shifted from the line connecting the centers of the Earth and the Moon.

Now the moon-brake keeps working on making days and nights longer. At the end of the story the period of the Earth daily rotation will get equal to that of the Moon orbiting and the impeding will cease. The Earth will forever remain turned to the Moon by the same side, like presently the Moon is. The increase of the day will affect the climate. The extended day on the sunny side of the globe will be opposed by the prolonged night on the rear. Cold air from the night side will rush to the warm hemisphere. Winds and dust-storms will break out... But the prospective is so far that the mankind will definitely find how to prevent these calamities.

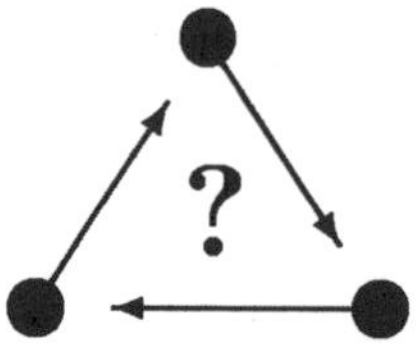

What is the effect of the moon-brake onto the duration of the lunar month (that is the period of orbiting of the Earth by the Moon)?

PART II

Saturday night physics

We've got so accustomed to our circumstances that don't even notice many wonders and think about actual causes behind them. Yet at closer sight one discovers a great many grounds for contemplation.

"When tossing pebbles into water, focus on the circles they produce, otherwise your tossing will be a mere time-frittering," — *wrote the great Koz'ma Prutkov*[b].

We shall try to convince you that even the most improbable phenomena in the world around us can be explained by the ordinary physics.

[b]Koz'ma Prutkov was the great Russian writer, poet and philosopher of nineteenth century. His selected works along with those by Alexander Pushkin and Mikhail Lomonosov may be found on the desk of any Russian scholar.

Chapter 8
Why the violin sings

The violin had no color, but sound it had.

N. Panchenko, The poem about a violin.

When an object moves through a medium, there always appear resistance forces trying to slow the object down. It will be the force of dry friction when a body slides mechanically along a rigid surface; in a liquid or in a gas that will be respectively the liquid friction (viscosity) and the aerodynamic resistance, and so on.

The interaction between a body and the surrounding medium is a rather complicated process leading usually to work-to-heat conversion of mechanical energy of the body. However, the reverse situation, when medium is in fact procuring body with energy is possible too. And this usually leads to some sort of oscillations. Just for example, the dry friction force between a pulled wardrobe (say you are moving) and the floor will brake it, slowing the motion; though the same force between the bow and string of a violin will make the string reverberate. As we will see later, the cause of vibration in the latter case is the dropping dependence of friction on the velocity of the motion. The vibration indeed occurs when friction decreases with augmenting velocity.

Let's illustrate generation of mechanical oscillations using as an example a concerto violin. The sound of violin is caused by the moving bow, right? It's impossible, of course, to explain here all the complicated phenomena involved in formation of a particular musical tone, yet let us try to understand in principle why the string starts vibrating when the bow is being smoothly pulled against it.

The friction force between bow and string is the dry friction. We can easily distinguish two different kinds of friction — friction of rest and sliding friction. The first acts between touching surfaces of two abutted bodies at rest with respect to each other; the second — when one body is actually sliding along the surface of the other.

As it's known, in the former case (no sliding), friction will balance an external force (being equal in magnitude and opposite in direction) up to a certain maximal value, called F^0_{fr}.

In turn, the sliding friction depends on the material and condition of the contacting surfaces, as well as on the relative velocity of the bodies. The latter circumstance we will discuss in more detail. The character of the relation between sliding friction and velocity varies for different bodies: often at first a drop in sliding friction is observed as velocity rises and, then, friction begins to go up too. Such dependence of magnitude of the dry friction force versus velocity is illustrated by the graph in Fig. 8.1. The friction force between the hair of the bow and a string behaves in this way too. When v, the relative velocity of bow and string, is zero, the friction between the bow and the string doesn't exceed F^0_{fr}. Then, for the descending wing of the curve, $0 < v < v_0$, any slight increase of the relative speed, by say Δv, leads to the corresponding decline of friction force and vice versa, when velocity is going down, the change of force will be positive (see Fig. 8.1). And, as we are going to show you in a minute, that's exactly due to this, not so evident at the first glance, feature that the energy of the string can grow on the expense of mechanical work done by the force of dry friction.

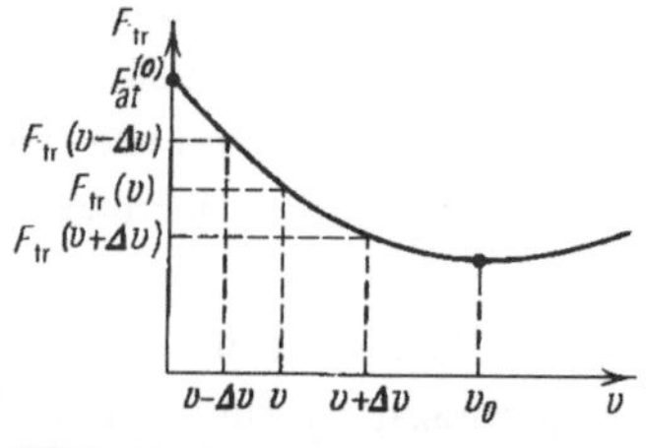

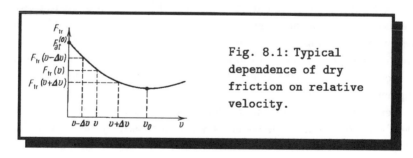

Fig. 8.1: Typical dependence of dry friction on relative velocity.

When the bow initially starts its motion, the string is getting drawn along with it, and the friction is compensated by the tension of the string,

Fig. 8.2. The resultant of the tension forces is proportional to the deviation x of string from the equilibrium:

$$F = 2T_0 \sin\alpha \approx \frac{4T_0}{l}x,$$

where l is the length of the string and T_0 is the tension force, which for small stretches, x, can be taken as constant. Thus, when the string is being pulled along with the bow, the force F is growing until it reaches the maximal value of friction, F^0_{fr}, and then the string begins sliding against the bow.

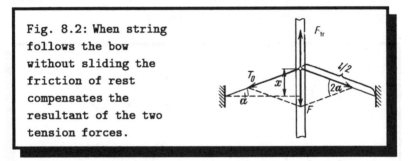

Fig. 8.2: When string follows the bow without sliding the friction of rest compensates the resultant of the two tension forces.

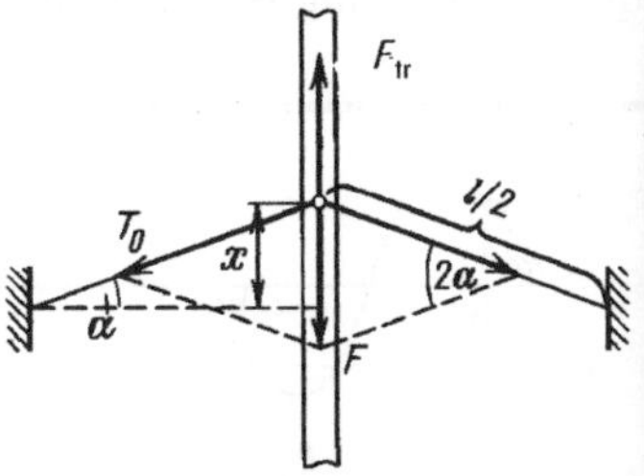

Let's, just for the simplicity sake, assume for now that at the beginning of slipping, the friction force drops abruptly from the maximum rest value F^0_{fr} down to a relatively weak force of sliding friction. In other words, we can approximately consider the slipping of the string as an almost free motion.

At the exact instant when the string from clinging to the bow takes off on sliding, its velocity is equal to the that of the bow, and, therefore, it keeps moving in the same direction. Yet, now the net tension force, not compensated by anything, will start slowing the motion of the string down. Consequently, at certain moment string's velocity will drop to zero, the string will stop and, then, will reverse its motion and go back, against the bow. Further, after a maximal swing to the other side, the string will again have to start moving in the same with the bow direction. Yet during all this time, the bow continues to move with the same constant velocity u and, therefore, at some point, the speeds of string and bow will match both in their magnitude and, this time, in the direction too. So the slipping between string and bow disappears and the friction force will balance that of the string tension again. Now, as the string approaches its neutral initial position, the tension force subsides causing the corresponding waning of the friction. And then, after the string passes the equilibrium position, everything happens again.

The ensuing graph for the string deviation versus time is shown in Fig. 8.3, *a*. The periodic motion of the string is composed of two different parts for each period. That is, for $0 < t < t_1$, the string moves at the same speed u with the bow so the deviation x is linearly proportional to the elapsed time ($\tan \alpha = u$). At t_1, the "take-off" occurs and, then, for the interval $t_1 < t < t_2$, the dependence of x on time becomes a sinusoid. At the instant t_2, when the tangent to the sinusoid has the same slope α as the starting linear piece of the curve (hence, the string and bow velocities are equal), the string is captured by the bow again.

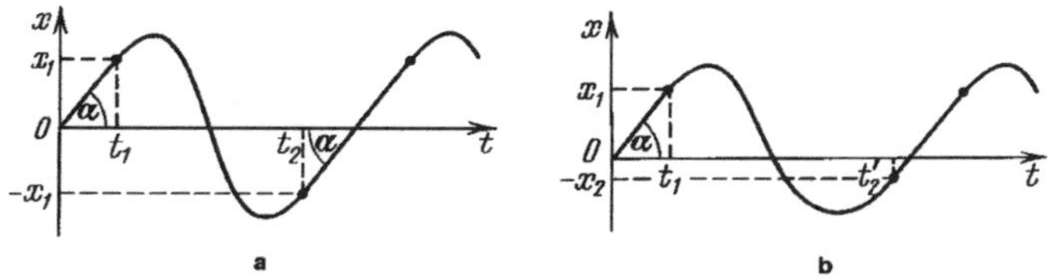

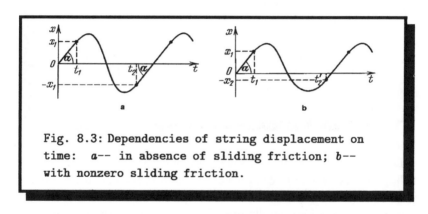

Fig. 8.3: Dependencies of string displacement on time: *a*-- in absence of sliding friction; *b*-- with nonzero sliding friction.

The Figure 8.3, *a* illustrates an ideal case, when there is no sliding friction force acting between the bow and the string and, consequently, there is no energy loss as the string moves freely. The total work performed by friction forces (in the intervals without slithering) in one complete cycle of oscillation of the string equals zero as well, because for negative x the mechanical work is negative — the friction force acts against the motion, whereas for $x > 0$ the work is the same in magnitude yet positive in sign.

Now let's try to figure out what happens if the sliding friction force is not counted as negligibly small any more. Well, it should cause energy loss for one thing. The graph for the string motion with sliding is presented in Fig. 8.3, *b*. For the positive x-values, the curve is actually steeper than for the negative ones. Hence, now the clinging of the string to the bow happens at a smaller in magnitude negative deviation ($-x_2$ in the picture) than the positive x_1 at which the string starts slipping off the bow at first: $x_2 < x_1$. Resulting is the positive mechanical work done by the friction forces in the intervals when the bow and the string go along together:

$$A = \frac{k\,(x_1^2 - x_2^2)}{2}$$

where $k = \frac{4T}{l}$ is the proportionality coefficient between the force of friction without sliding that draws the string off the equilibrium position and the string swing [a].

This positive fraction of the total work does indeed compensate the energy losses due to the sliding friction and makes the string oscillate without damping.

Generally speaking, to replenish the energy, it's not at all necessary for the string to keep clinging to the bow. It's enough if their relative velocity v stays within the descending part of the dependency between the sliding friction and the relative speed of the bow and string (look back at Fig. 8.1). Now let's take a closer look at the vibration of the string in this case.

Suppose the bow is being pulled with some constant speed u, and the string is driven away from its neutral equilibrium position by x_0 so that the net tension force $F(x_0)$ is again compensated by the sliding friction force $F_{fr}(u)$. If, by chance, the string deviates in the direction of the motion of the bow, their relative velocity shall decrease causing friction to rise (notice that we are speaking of the "dropping" part of the $F_{fr}(u)$ curve!) which, in turn, makes the string stretch even more. As the string stretches further, at some point, the elastic force will necessarily exceed the friction (remember that the vector sum of tensions is directly proportional to deviation of the string from the neutral position, whereas friction is limited by F_{fr}^0), and the string will reverse its motion and go back in opposite direction. Then, continuing to move, the string will successively pass the equilibrium state, go on further, stop in the utmost position on the other side and, then, will repeat everything again... Thus oscillations will be amplified.

It's important to notice that the described oscillations once started will proceed without dumping. Indeed, when the string moves with a velocity Δv and $u > \Delta v > 0$ in the direction of the bow, then the friction performs positive work. On the other hand, when going back the work of friction will

[a]Remember that at the linear piece of the curve, Fig. 8.3, the force of friction is equal in magnitude to the resultant of the tension forces, Fig. 8.2

be negative. But the relative velocity $v_1 = u - \Delta v$ in the former case is less than that, $v_2 = u + \Delta v$, in the latter one, and on the contrary, the friction, $F_{\text{fr}}(u - \Delta v)$, will be greater in the first situation than that, $F_{\text{fr}}(u + \Delta v)$, in the second. Thus, the positive mechanical work done by friction when the string and bow are moving together surpasses the negative one performed when the string moves back, resulting in the positive net work during the vibration cycle. Consequently, the amplitude of vibrations increases with each successive oscillation. And it keeps going up until it reaches a certain limit. If $v > v_0$ so that the relative velocity of bow and string v finally goes out of the descending part of the graph $F_{\text{fr}}(v)$ (Fig. 8.1), then the negative work of friction can overcome the positive one, forcing the amplitude of oscillations to wane.

As the result, a stable vibration with an equilibrium amplitude will be finally attained, for which the total work done by friction will be exactly equal to zero. (To be precise, the positive work during the cycle compensates the energy loss due to the air resistance, nonelastic character of deformation *etc.*). These steady oscillations of the violin string will proceed without damping.

It's quite common that sound vibrations are excited when one body moves along the surface of another: dry friction in a door hinge causes it to screech; and so do our shoes, floor tiles and so on. You can produce screeching by just pressing and pulling your finger along, say, a smooth and firm enough surface[b]. And the phenomena which occur in these examples may be very similar to the excitation of vibrations of the violin string. At first — there is no sliding, then, an elastic deformation develops, up to the point when the "take-off" happens and "their majesty" oscillations commence. And once having started, they don't subside abruptly rather continuing without significant damping because of that described "dropping" character of the friction forces procuring the required energy due to their mechanical work.

If the dependence of friction force on relative velocity of moving surfaces changes its character, the screech goes away. Everyone knows, for instance, that you can simply lubricate the surfaces to get rid of the irritating screak. And the physical reason behind it is just trivially that the liquid friction is proportional to velocity (in case of low velocities) and, hence, the conditions required to induce and then sustain oscillations disappear when

[b]Chapter 9, "The chiming and silent goblets" gives a less trifle example of the kind.

you substitute dry friction with the liquid one. Inversely, when vibrations are desirable, participating surfaces are often treated in a special way so to reach a sharper decrease of friction force with increasing velocity. For instance, for this exact reason they apply rosin on the violin bows.

No surprise that understanding the laws of friction often helps solving different practical and industrial problems. For example, while machining a metal piece, undesired vibrations of the cutter can develop. These vibrations are caused by the force of dry friction between the tool and metal shaving slithering along its surface, Fig. 8.4, for the friction force versus speed relation for high quality types of steels may have that familiar "dropping" character. Which is, as we know by now, is the principal condition for exiting oscillations. A common way to preclude such vibration (which can turn out to be quite detrimental for both the cutter and the piece in work) will be to use along with naive lubrication a special sharpening of the cutter, basically to hone it a correct angle, so that the slithering wouldn't occur and there would be no reason for oscillations.

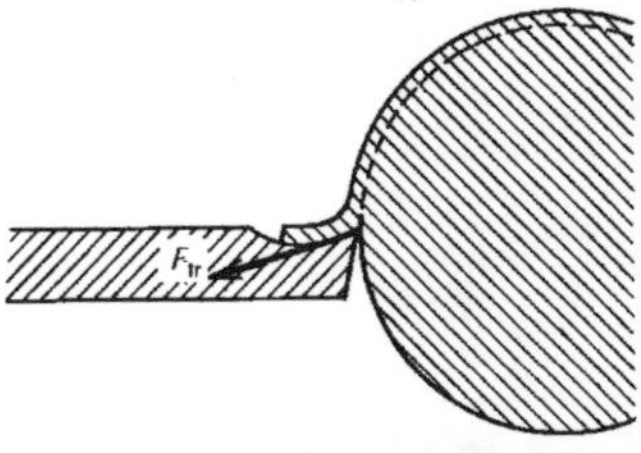

Fig. 8.4: Vibrations of cutter of machine tool may be eliminated by a right choice of the sharpening angle.

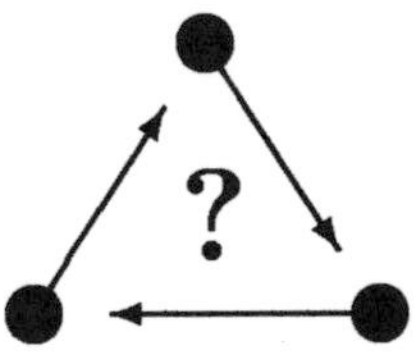

Can you describe (or even write a formula for) the motion of string subject to constant sliding friction?

Chapter 9

The chiming and silent goblets

> The carriage resembled an open shell made of glittering crystal; its two large wheels seemed to be built of the same material. When they were turning, they produced marvelous sounds: Full, yet still growing and approaching, these chords reminded the tones of glass harmonica, yet of amazing strength and power.
>
> *E. T. A. Hoffmann,* Klein Zaches.

It's not a novel idea that one can make simple wine-glasses sing. However, it turns out that there is a very peculiar way to do so. How peculiar? Well, you judge for yourself.

If you dip your finger in water and start circling it carefully along the edge of a glass, wetting the rim constantly, at first, it makes a rather screeching sound, but then, after water has covered the glass edge thoroughly and uniformly, the tone should turn into something more melodic. By varying the pressing force of the finger, one can easily change the pitch of the produced sound. The height of the pitch will also depend in this case on the size of the glass and thickness of its walls[a].

Notice, by the way, that not every single glass is capable of making those chiming pleasing tunes, so the search for a suitable one may turn out to be a quite cumbersome affair and take a while. The best "singers" turn out to be very thin-walled goblets, those having shape of the paraboloid of revolution, with a long slim stem. Another critical parameter determining

[a]The mechanism of exciting sound is the same as in the bow-instruments, see Chapter 8, "Why the violin sings."

the resonant tone of the glass is the level of liquid in it: generally, the fuller is the glass the lower its pitch is. When the water level passes the midline of the glass, waves will develop on the surface of liquid, because of the wall shaking. The maximal disturbance marks the position of the finger inducing the sound at the moment.

A famous American scientist (as well as one of the greatest statesmen in the history of his country — rare yet proven possible, at least back in those days, combination) Benjamin Franklin, who is mostly known for his experiments with atmospheric electricity[b], had employed the discussed in the previous paragraph phenomenon to create a peculiar musical instrument, very similar to that described in Hoffmann's "??". That was a series of perfectly polished glass cups, each with a drilled orifice in the middle, arranged equidistantly on the same axle. There was also a pedal underneath of the box where the cups on the axle were situated, kind of like in an old-fashioned sewing machine, to make the axle rotate. And just by simple touching with wetted fingers, one could change the tone of the system from a sound forte down to meager whistling. Now it is hard to believe, but the people who had heard this "goblet organ" playing assured that the harmony of its sounds was acting amazingly appeasing on both the listeners and the performer. In 1763 Franklin had given his own instrument as a present to an Englishwoman, Ms. Davis. She was demonstrating it for several years traveling around Europe, and then, the famous instrument had disappeared without a trace. Probably memories of that true story had reached E. T. A. Hoffmann, who was a talented musician himself, and were used in his "??".

And since we are talking glasses, it seems also worth mentioning a rather interesting fact that, iconoclastic though it may sound, it's not actually accepted as a proper etiquette rule to clink the champagne glasses. Really. And the deal here is that for some, of course, purely physical reason goblets filled with champagne or any fizzy carbonated drink make, when clinked, an inexpressive muffled than sound. So what's the matter here? Why don't goblets filled with champagne ring?

The physical phenomenon responsible for melodic ear pleasing tingle we hear clinking the glasses are the high frequency (that is, in the range $10 - 20\,KHz$) sound and even ultrasound (higher than $20\,KHz$) waves excited in the resonators, which our glasses are in this case. When we

[b]B. Franklin, (1709–1790), American public official, writer and scientist.

clink either empty glasses or glasses filled with a noncarbonated beverage, these oscillations once having been induced keep ringing for a rather long period of time. That, on the other hand, automatically suggests as a likely cause for the muffled sound produced by goblets filled with champagne those tiny pinching bubbles of carbon dioxide, stampeding from an opened bottle in their delectable effervescent rush. They may well lead to a strong scattering of those short wave sound oscillations in the goblet, in the way similar, for example, to that taking place in the atmosphere. Remember that fluctuations of molecular density cause scattering of sun light in the short wave part of the spectrum (see Chapter 4, "In the blue").

Even for the highest frequencies perceptible by human ear ($\nu \sim 20\,KHz$), the wave length of sound in water, $\lambda = c/\nu \sim 10\,cm$ R ($c = 1450\,m/s$ is velocity of sound in water), is considerably larger than the size of the CO_2 bubbles in champagne (say, about $1\,mm$), and, consequently, the latter seem to be quite legitimate candidates to cause the Rayleigh-type scattering of sound. Yet, let's look at the problem a bit closer. What does for instance our estimate for $\lambda_{\min}$ really mean? Just for simplicity, let's forget for now about the complicated shape of a real goblet and think of a rectangular box, with a plane one-dimensional expansion-compression sound wave in it. We can write for the excess air pressure in the box:

$$P_e\,(x,\,t) = P_0\,\cos\left(\frac{2\pi\,x}{\lambda} - \omega\,t\right), \tag{9.1}$$

where P_0 is the amplitude of pressure oscillations, ω is the sound frequency, λ is the corresponding wave length and x is the coordinate along the axis of propagation.

Since even the minimal value of λ is well over the glass dimensions, $x \ll \lambda_{\min}$ the function $P_e(x,\,t_0)$ (called also the ***pressure field*** of the sound waves) at any given instant t_0 varies just slightly over the whole volume of the glass. Thus, the first term in (9.1) is vanishing small, and the "space-time" distribution of the excess pressure is mainly determined by the second term in the argument of the cosine. This actually shows that because of the negligible value of $x \ll \lambda$ there is an almost uniform yet rapidly changing excess pressure field inside the glass:

$$P_e\,(t) = P_0\,\cos\omega\,t, \tag{9.2}$$

Note the difference, by the way, between the shown pressure field evolution (9.2) and conventional standing waves in a rigid box of length l. The resonance condition for those would read: $l = \frac{n\lambda}{2}$, where $n = 1, 2, \ldots$ Such sound waves simply wouldn't fit in the glass. However the walls of a real glass are elastic take part in the oscillations of the contents. The vibrations of the walls transfer sound to the ambient air making it audible[c].

Thus, we can express the total pressure of liquid in the glass as the sum of $P_e(t)$ and the atmospheric pressure:

$$P_e(t) = P_{\text{atm}} + P_0 \cos\omega t,$$

and now are just one step away from understanding the true reason behind the observed fast damping of the ringing sound in the goblets with champagne. And the answer is hiding in the fact that liquid saturated with gas turns out to be a so-called ***nonlinear acoustic medium***. This piece of "scientific" vernacular means in reality the following. It's known that the solubility of gas in liquid depends on pressure, — the higher is the pressure the greater is the gas volume soluble in unit volume of liquid. But, as we have established already, the pressure field in the glass where sound oscillations are exited varies. At the moments when the pressure in the liquid drops below the atmospheric one, the outgassing consequently increases. Of course, the released gas bubbles distort the simple harmonic time-dependence of the pressure and in this particular sense one calls the gas saturated liquid a nonlinear acoustic medium[d]. The outgassing inevitably takes energy from the sound oscillations making them wane much faster. At first after the glasses have been clinked, there are all kinds of sound frequencies exited in them; then, however, due to the sketched above mechanism, the high pitch-modes will subside far quicker than the low-pitch ones resulting in the muffled thump rather than the pure high-pitched ringing melody of crystal.

Yet, it turns out, on the other hand, that gas bubbles in liquid not only do damp sound waves but can also in certain circumstances generate them. Indeed, it has been found recently, for instance, that sound oscillations can be excited by wee air bubbles in water when subjected to high power pulse laser radiation. The effect is caused by the impact of the laser beam upon the bubble surface, from which, because of the total internal reflection, the beam can bounce back. After such a "hit", the bubble quivers for some time

[c]The crystal voice of empty wine-glasses proves that what we usually enjoy is not the solo of liquid contents but rather its duet with the splendid vessel. —*A. A.*

[d]Remember the nonlinear sound distortions which are the nightmare of Hi-Fi fans.

(until the vibration is damped) exciting sound waves in the surrounding medium. Let's now try to evaluate the frequency of these oscillations.

There are lots of important phenomena in nature, which, whatever different they may seem, could be described by the same equation, the equation of harmonic oscillator. These are different kinds of oscillations — a weight bouncing on a spring, atoms vibrating in molecules and crystals, electrical charge flowing back and forth from plate to plate of the capacitor in LC contour and well many others. The imperative physical feature uniting all the above examples is the presence of a "restoring force" that depends linearly on the displacement and always tends to bring the system back to the equilibrium, once it has been driven away from that by some external perturbation. And the gaseous bubbles vibrating in liquid are just one more example of such oscillatory systems. Hence, we can try to use the well-known relation for the vibration frequency of a mass on a spring to estimate the typical frequency of oscillations of the bubbles. Of course to do so, we have to figure out what would be in this case the "coefficient of elastic force".[e]

The first candidate for this part could be the surface tension σ of the liquid: $k_1 \sim \sigma$; it has the desired dimension (N/m) at least. Instead of mass of the weight, it seems reasonable to put into the formula for the oscillator's natural frequency the mass of liquid involved in the bubble oscillations. Clearly the sought mass should be close to the volume of the bubble times the liquid's density: $m \sim \rho r_0^3$. So for the natural frequency of the oscillations of the bubble, one can write the following expression:

$$\nu_1 \sim \sqrt{\frac{k_1}{m}} \sim \frac{\sigma^{\frac{1}{2}}}{\rho^{\frac{1}{2}} r_0^{\frac{3}{2}}}.$$

However it turns out to be not the only possible solution. We haven't yet utilized in any way another important parameter, that is the air pressure inside our bubble, P_0. When multiplied by the radius of the bubble, it gives the same (N/m) dimension of the "elasticity coefficient" too, $k_2 \sim P_0 r_0$. And after having plugged this new coefficient in the same relation for the

[e]Coefficient of elastic force of a spring k characterizes the proportionality between the value of the restoring force F and the displacement x from the equilibrium:

$$F = -kx.$$

natural frequency of oscillator, we obtain an entirely different value for the frequency of bubble's vibration:

$$\nu_2 \sim \sqrt{\frac{k_2}{m}} \sim \frac{P_0^{\frac{1}{2}}}{\rho^{\frac{1}{2}} r_0}.$$

Which of these two values is the true one? Surprising though it may sound, yet both of them are correct, actually. They just correspond to two different types of oscillations of the air bubble. That's it. The first one represents oscillations occurring when the bubble was originally squashed, say by the laser impulse. In such a motion, the shape and, therefore, the surface area of the bubble are constantly changing, yet its volume remains the same. The resulting "restoring force", in this process, is determined by the surface tension[f]. The second type, on the other hand, takes place when the bubble had been squeezed uniformly from all directions, and then let go. In this case, it starts throbbing already due to the pressure forces. And the second of the found frequencies indeed describes the radial oscillations of this kind.

Because of the obvious asymmetrical character of the laser beam impact, the sound waves produced by the bubbles are likely to belong to the first of the two considered types. Further, if, for instance, the size of bubbles is known, one could determine the type of vibration from the frequency of the sound generated by the bubbles. In the discussed experiments, this frequency was found to be $3 \cdot 10^4\, Hz$. Unfortunately dimensions of the tiny air bubbles in water are hard to measure with the sufficient degree of accuracy. It's clear though that they should be of the order of some fractions of a millimeter. After plugging $\nu_0 = 3 \cdot 10^4\, Hz$, $\sigma = 0.07\, N/m$, $P_0 = 10^5\, Pa$, $\rho = 10^3\, kg/m^3$, in the corresponding formula, one finds the characteristic dimensions of the generating sound bubbles, for both types

[f]Notice that a whole bunch of different kinds of oscillations, in which the bubble's volume doesn't change is possible. Those vary from trivial alternated squashing and squeezing the bubble in various directions to much more outlandish transformations when the bubble turns into something like, say, doughnut. The frequencies of such oscillations may variegate somehow quantitatively yet remain of the same order of magnitude equal to

$$\nu_1 \sim \frac{\sigma^{\frac{1}{2}}}{\rho^{\frac{1}{2}} r_0^{\frac{3}{2}}}.$$

of oscillations:

$$r_1 \sim \frac{\sigma^{\frac{1}{3}}}{\rho^{\frac{1}{3}}\,\nu_0^{\frac{2}{3}}} = 0.05\,mm;$$

$$r_2 \sim \frac{P_0^{\frac{1}{2}}}{\rho^{\frac{1}{2}}\,\nu_0} = 0.3\,mm.$$

So, as it turns out, the size of the bubbles doesn't differ much, at least not enough for us to distinguish what type of oscillations was generated in the described experiment. As a sort of confirmation of the correctness of our reasoning (which used, mainly, the dimensional arguments), we can accept the fact that the estimated radii of the bubbles are in complete agreement with what we would expect from our day-to-day observations.

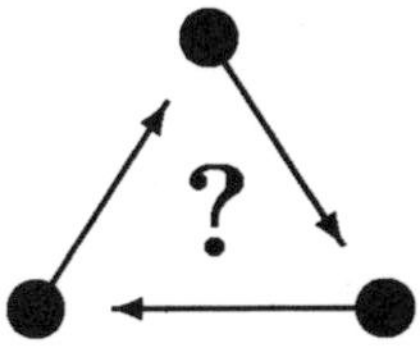

Do you have any ideas why in the glass of champagne the high-pitch overtones die out far quicker than the principal mode?

Chapter 10
The bubble and the droplet

Pervasive and amazingly various are numerous guises of surface tension in the natural and technological world around us. It gathers water into droplets, because of it one can blow up a shimmering with rainbow soap-bubble, or write with an ordinary pen. Surface tension plays also a significant role in the physiology of human body. It's been utilized in the space technology too. And why, after all, does the surface of liquid behave in the way it does, like a stretched elastic membrane?

The molecules in the narrow layer, real close to the liquid surface, could be considered as "dwelling" in very special circumstances. They happen to have neighbors, the identical to them molecules, only on one side, whereas the "inner" denizens are completely surrounded by their twin looking (and acting) relatives.

Because of attractive interaction between the closely lying molecules the potential energy of each of them is negative. Its absolute value, on the other hand, could in first approximation be premised as proportional to the number of the nearest neighbors. Then it's clear that surface molecules, each having fewer neighbors right next to it, must have a higher potential energy than the ones in the volume of the liquid. Another factor raising the potential energy of the molecules in the surface layer is that the concentration of molecules in the liquid decreases near the surface.

Of course, molecules of liquid are in their incessant thermal motion — some of them dive inside, leaving the surface, and others go up to take their place. However, one can always speak of the average surplus potential energy of the surface layer.

The reasoning above shows that in order to extract a molecule from inside the liquid up to the surface, external forces must perform a positive work. Quantitatively this work is expressed in by the surface tension σ, which is equal to the additional potential energy of molecules occupying a unit surface area (compared to the potential energy these molecules would have if they remained inside).

We know that the most stable, among all possible states of a system, is that with the lowest potential energy. In particular liquid will always try to assume the shape corresponding to the minimal surface energy for the given conditions. This is the origin of surface tension, which tries actually to always shrink the surface of liquid.

10.1 Soap-bubbles

Paraphrasing the great English physicist Lord Kelvin[a], you can simply blow up a soap-bubble, stare at it, study it all your life long and still be able to extract more and more lessons of physics. For instance, the soap film is an excellent object for exploring various effects of surface tension.

Gravitational forces do not play any noticeable part in the considered case, for the soap film is very thin and, therefore, its mass is negligible. So the protagonist here will be the surface tension force which, as we've just shown, will try to make the surface area of the film as small as possible, within the given circumstances, of course.

But why necessarily the soap films? Why can not one, for example, study, say, films of the distilled water? Especially considering the fact that its surface tension is several times that of the aqueous soap solution (just a fancier name for the soapy water).

It turns out that the answer does not depend so much on the value of the surface tension coefficient, but lies rather in the structure of the soap film itself. Indeed, any soap is abundant with so called surface-active agents *(surfactants)*, pretty long organic molecules with their two ends having completely opposite affinity to the water: that is, while one end (called "the head") clings to water avidly, the other one ("the tail") stays completely water indifferent. This" leads to the rather complex structure of the soap film in which the soapy solution is armored by a fence made of

[a]See page 53.

those densely packed highly oriented layers of the surface-active agents[b], Fig. 10.1.

Fig. 10.1: Stability of soap film is guaranteed by presence of surface-active organic molecules.

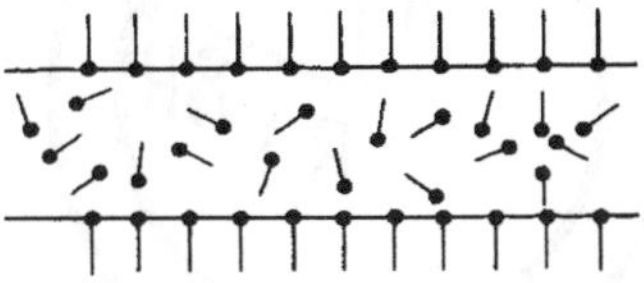

But for a moment, let's go back to our soap-bubbles. Most of us not only just marveled these gorgeous creations of nature at one occasion or another, but were making them ourselves. They are so perfectly spherical in their shape and can hover in the air for so long, before rupturing finally against an obstacle. The pressure inside them appears to be higher than the atmospheric one. This additional pressure is due to the fact that the soap film of the bubble, attempts to minimize the surface area and gives to the air inside an extra squeeze. Moreover, the smaller the bubble radius R is, the higher is the additional pressure inside. Now we shall try to find the magnitude of this addition, ΔP_{sph}.

Let's conduct a so-called mental experiment. Suppose that the surface tension of the film of the bubble drops a tiny bit, and its radius increases, consequently, by a certain value, $\delta R \ll R$, Fig. 10.2. This, in turn, causes the following increase of the external surface area:

$$\delta S = 4\pi\,(R + \delta R)^2 - 4\pi\,R^2 \approx 8\pi\,R\,\delta R;$$

($S = 4\pi\,R^2$ stands for the surface area of the sphere). And, therefore, for the incremental surface energy, one can write:

$$\delta E = \sigma\,(2\,\delta S) = 16\pi\,\sigma\,R\,\delta R, \tag{10.1}$$

(since δE is proportional to the tiny δR, the surface tension coefficient σ can be assumed to be constant).

[b] Surfactants are mainly used in order to reduce the surface tension and improve wetting properties of detergents. In the mean time they help to stabilize the film and prolong the lifetime of soap-bubbles. —*A. A.*

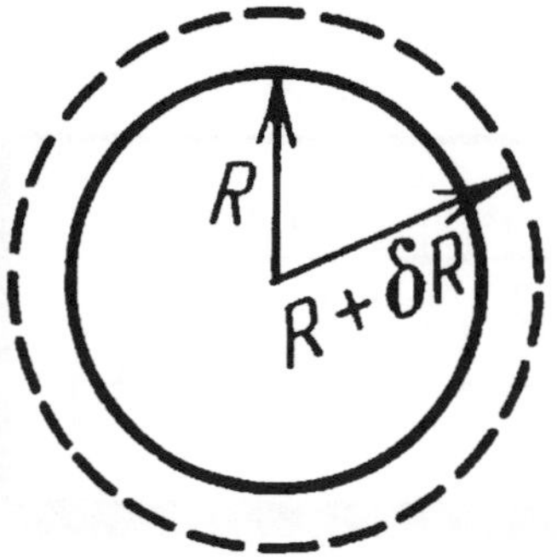

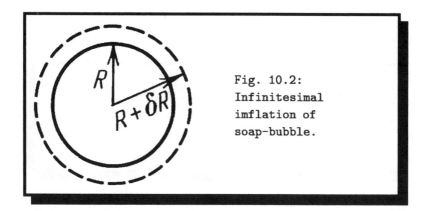

Fig. 10.2: Infinitesimal imflation of soap-bubble.

By the way, notice the extra "2" factor showing in (10.1), although absent from the original definition of surface energy. That is because now we have taken into account both surfaces of the bubble, its internal surface as well as the external one; when the bubble radius grows by δR each of its surfaces stretches by additional $8\pi\, R\,\delta R$.

This fictitious increment of surface energy thanks to mechanical work of the compressed air trapped inside the bubble. The pressure in the bubble remains almost the same when its volume grows by a small amount δV so one can equate this work to the enhancement of surface:

$$\delta A_{air} = \Delta P_{sph}\, \delta V = \delta E.$$

The volume change here, on the other hand, is equal to the volume of the thin-walled spherical shell, Fig. 10.2:

$$\delta V = \frac{4\pi}{3}(R+\delta R)^3 - \frac{4\pi}{3} R^3 \approx 4\pi\, R^2\, \delta R,$$

entailing,

$$\delta E = 4\pi\, R^2\, \Delta P_{sph}\, \delta R.$$

Now compare this expression to the earlier established formula (10.1). This gives for the additional pressure inside the spherical soap-bubble that balances the surface tension forces:

$$\delta P_{sph} = \frac{4\sigma}{R} = \frac{4\sigma'}{R} = 2\sigma'\, \rho; \tag{10.2}$$

(We denoted by $\sigma' = 2\sigma$ the doubled coefficient of surface tension of the liquid).

Obviously, in case of a single curved surface (for instance, that of a spherical droplet), this additional pressure would be $\delta P_{sph} = 2\sigma / R$. This relation is called the ***Laplace formula***[c]. The reciprocal of radius is conventionally called the ***curvature*** of the sphere: $\rho = 1/R$.

Thus we have arrived to the important conclusion that the incremental pressure is proportional to the sphere's curvature. Yet sphere is not the only shape a soap-bubble can take. Indeed, having placed the bubble between two hoops[d], for example, one can easily stretch it in a cylinder crowned by round spherical "caps", on its top and bottom Fig. 10.3.

Fig. 10.3: With the help of wire frames you may make a cylindrical soap-bubble.

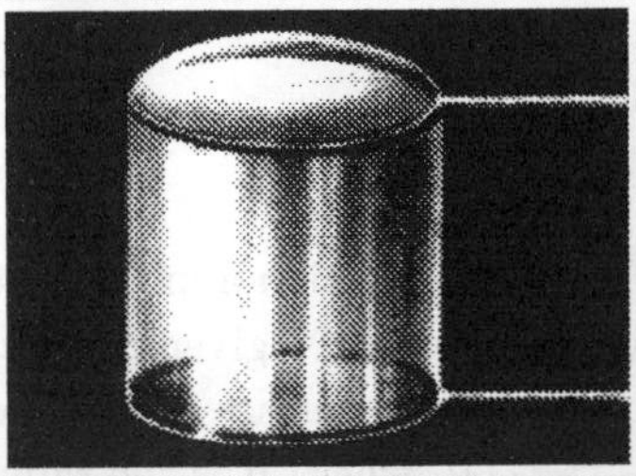

What will the value of the additional pressure be for such an "unorthodox" bubble? It's clear that the curvature[e] of cylindrical surface varies in different directions: it is zero along the generating line (for cylinders that is straight), however for the cross section perpendicular to the axis, the curvature equals $1 / R$, where R is the radius of the cylinder. — Well, then

[c] Pierre-Simon Laplace, (1749 - 1827). A great French mathematician, one of that splendid constellation of French mathematicians, contemporaries of the Great French Revolution — J. L. Lagrange, L. Carnot, A. M. Legendre, G. Monge, *etc.* Laplace became the best known for his contribution to the theory of probability and celestial mechanics, as well as for the famous quote from Napoleon that he "carried the spirit of the infinitely small into the management affairs", when the scholar had miserably failed in his fast-ended assignment as "Minister of the Interior".

[d] Before touching a bubble you must dip the hoops into the soap solution. —*A. A.*

[e] What is the curvature of a plane (two-dimensional) curve? For a circumference it is defined in the same way as for the sphere: $\rho = 1 / R$, where R is the radius. Each tiny piece of any other curve can, just the same, be considered as an arc of certain radius. The reciprocal of this radius is called the curvature of the plane curve at the given point.

what value of ρ should we substitute into the previously derived formula? It turns out that the difference of pressures on the two opposite sides of an arbitrary surface is defined by the average curvature of the latter. Let's try to figure out what will it be for our right cylinder?

First, erect a normal[f] to the cylinder surface at a point A, and then construct a set of planes passing through the normal. The resulting cross sections of the cylinder by these planes, (called the ***normal sections***), can be either a circle, or an ellipse, or even degenerate into two parallel straight lines, Fig. 10.4. Surely, their curvature at the given point is different: it is maximal for the circle and minimal (nil actually) for the longtitudinal section. The average curvature is defined then as the half sum of minimal and maximal values of the curvature of the normal sections at this point:

$$\bar{\rho} = \frac{\rho_{max} + \rho_{min}}{2}.$$

The given definition applies not only to the cylinder, — in principle, the average curvature at a given point can always be calculated in this manner.

For the lateral surface of a cylinder, its maximal curvature at any point is $\rho_{max} = 1/R$, where R is the radius of the cylinder, whereas the minimal value $\rho_{min} = 0$. Thus, the average curvature of the cylinder is $\bar{\rho} = 1/2R$, and the additional pressure inside the cylindrical bubble is:

$$\Delta P_{cyl} = \frac{\sigma'}{R}.$$

So it turns out that the additional pressure in the cylindrical bubble is equal to that in the spherical bubble of the doubled radius.That makes the radii of the spherical caps of such a cylindrical bubble twice the radius of the cylinder itself. Hence the caps are just spherical segments rather than the full hemispheres.

And what would happen if one eliminates the additional pressure in the bubble completely by, say, pricking its caps? The first solution popping to head would be that, since there is no any additional pressure, the surface shouldn't have any curvature at all. Yet surprisingly, the walls of the cylinder are actually bending inside taking the shape of a ***catenoid*** (from the Latin *catena* for "chain"). This shape can be generated by rotating the so

[f]That is the line perpendicular to the tangential plane at the point A. —*D. Z.*

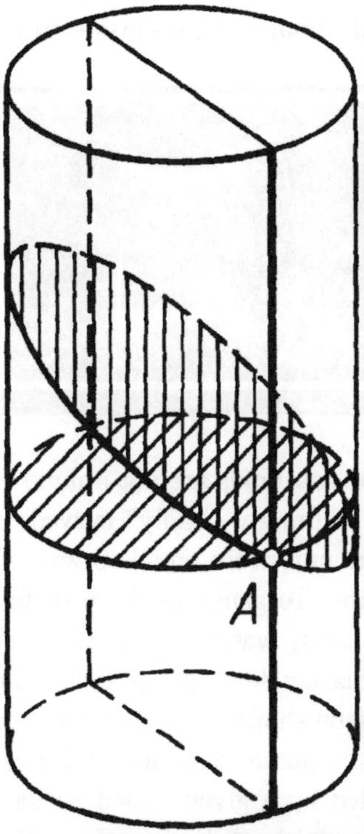

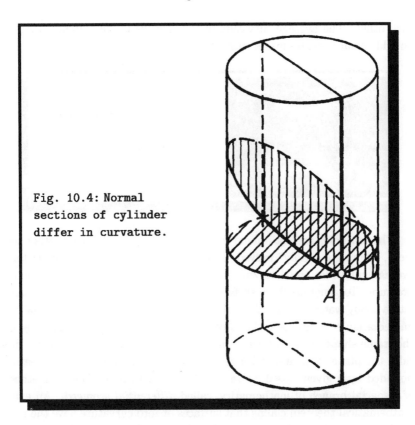

Fig. 10.4: Normal sections of cylinder differ in curvature.

called *catenary line,* around its X-axis[g]. So what's the matter here?

Let's examine this surface closer, see Fig. 10.5. It's easy to notice that its narrowest (waist) portion, also called *saddle,* is concave as well as convex. Its section across the rotation axis is obviously a circumference, on the other hand, dissecting along the axis gives, by definition, the catenary. The inward curvature should raise pressure inside the bubble, but the opposite curvature would lower it. (Pressure under a concave surface is higher than the pressure above it.) In the case of catenoid , the two curvatures are equal in magnitude but have the opposite directions, therefore, negating

[g]Catenary is the curve formed by a perfectly flexible uniform chain suspended by the endpoints. The form of the curve (up to similarity transformations $a \to \alpha a$) is given by the equation:

$$y = \frac{a}{2}(e^{\frac{x}{a}} + e^{-\frac{x}{a}}).$$

—*D. Z.*

each other. The average curvature of this surface is zero. Hence, there is no additional pressure inside such a bubble.

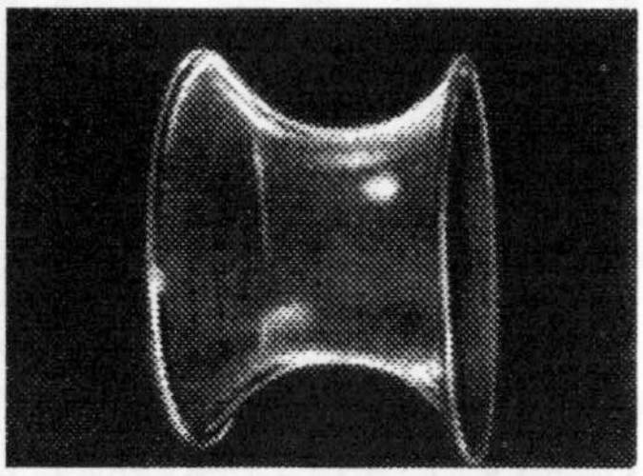

Fig. 10.5: Left to itself the soap film takes the form of catenoid. This surface has zero average curvature.

The catenoids are not unique though, and there is a bunch of other surfaces, seemingly "badly" curved in all possible directions, yet having their average curvature equal to zip, and consequently, not exerting any extra pressure. To generate these surfaces, it's enough to immerse a wire frame into a soapy water. While lifting the frame back from the solution, one can immediately see various surfaces of zero curvature, formed depending on the frame shape. However, catenoid is the only surface of revolution[h] (besides the plane, of course) with zero curvature. Surfaces of zero curvature bounded by a given closed curve may be found with the help of methods of a special branch of mathematics, called ***differential geometry***. An exact mathematical theorem claims that surfaces of zero curvature have the minimal area among all the surfaces with the same boundary; the statement which seems pretty natural and obvious for us now.

A plethora of combined together soap-bubbles makes up froth. In spite of the seeming disorder there is an indisputable rule held in the embroidery of soap films in the foam: the films intersect one another only at equal angles, Fig. 10.6. Indeed, look for example at the two joint bubbles partitioned by the common wall, Fig. 10.7. The additional (with respect to atmospheric) pressures inside the bubbles will be different. According to the Laplace formula, (10.2):

$$\Delta P_1 = \frac{2\sigma'}{R_1}, \qquad \Delta P_2 = \frac{2\sigma'}{R_2}.$$

So the common wall of must be bent in order to compensate the pressure

[h]That is the surface which may be generated by rotation of a curve.

difference in the bubbles. Its radius of curvature, hence, is determined by the expression:

$$\frac{2\sigma'}{R_3} = \frac{2\sigma'}{R_2} - \frac{2\sigma'}{R_1},$$

which gives after regrouping

$$R_3 = \frac{R_1 R_2}{R_1 - R_2}.$$

Again, Fig. 10.7 depicts a cross section of these two bubbles by a plane passing through their centers. The points A and B here mark the intersections of the plane of the picture with the circumference where the bubbles are touching. At any point of this circumference there are three films coming together. As long as their surface tension is the same the tension forces can balance each other only if the angles between the crossing surfaces are the same. Therefore, each of them is equal to 120°.

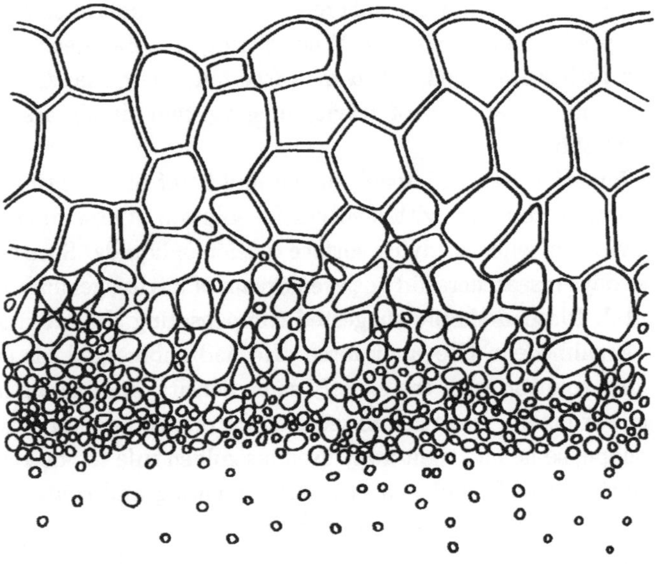

Fig. 10.6: Section of soap froth shows that joining films make equal angles.

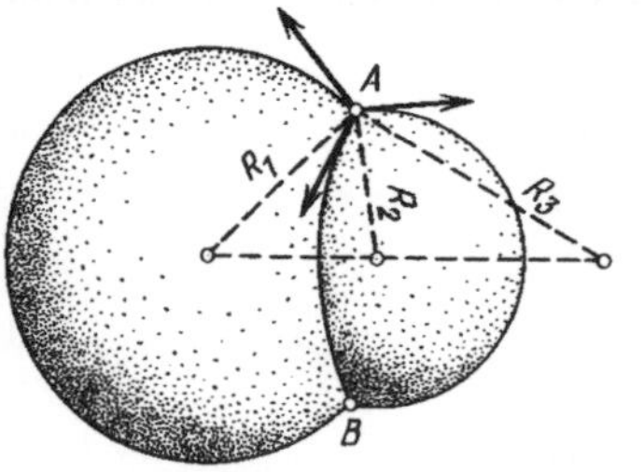

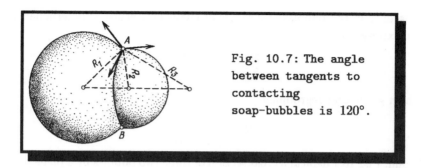

Fig. 10.7: The angle between tangents to contacting soap-bubbles is 120°.

10.2 On different kinds of droplets.

The shapes of droplets. Here things are becoming a bit more complicated. Now the surface tension which, as always, tries to minimize the area of the surface is counteracted by other forces. For example, a liquid droplet almost never is spherical, although it is a sphere that has the smallest surface area among all shapes for a given volume. When sitting on a flat steady surface, droplets look rather squashed; when in a free fall, their shapes are even more complex; only in absence of gravity in space they finally assume the form of perfect spheres.

The Belgian scientist J. Plateau[i], in the middle of nineteenth century, was the first one to come up with a successful solution of how to eliminate effects of gravitation when studying surface tension of liquids. Sure enough, back in those days researchers did not even dream of having genuine weightlessness, and J. Plateau simply suggested compensating the gravitational forces with Archimedean buoyancy force. He had submerged his subject liquid (oil) in a solution with the exactly same density, and, as his biographer tells us, was utterly surprised to see that the oil droplet had developed a spherical shape; so he had right away used his golden rule to "become surprised in the right time", and was then experimenting and contemplating upon this peculiar phenomenon for a long while.

He had used his method to study a variety of entailing effects. For instance, he meticulously investigated the process of droplet formation at

[i]Joseph Antoine Ferdinand Plateau, (1801–1883), Belgian physicist; Works in physiological optics, molecular physics, surface tension. Plateau was the first to put forward the idea of stroboscope.

the end of a tube.

Normally, no matter what slow a droplet is being made, it separates from the tip of the tube so fast that human eye can't follow the details of this event. So Plateau had to dip the tip of the tube he was using into a liquid, with the density just slightly less than that of the droplets themselves. The gravitational force influence was, by doing so, substantially diminished and as a result, really large droplets could be formed and the process of their taking off from the end of the tube could be clearly seen.

In Fig 10.8 you see the different stages droplets undergo in their formation and separation (of course, these pictures were taken using the modern high-speed filming technique already). Let's try now to explain the observed sequence. During the slow growth stage, the droplet can be accounted as being in equilibrium at each particular instant. For a given volume, the droplet shape is determined by the condition of minimal sum of its surface and potential energies, the latter of which is the result of the gravitational forces, of course. The surface tension is trying to shape droplets spherically, whereas the gravitation, on the contrary, tends to situate the droplet's center of masses as low as possible. The interplay of these two yields the resulting vertically stretched form (the first shot).

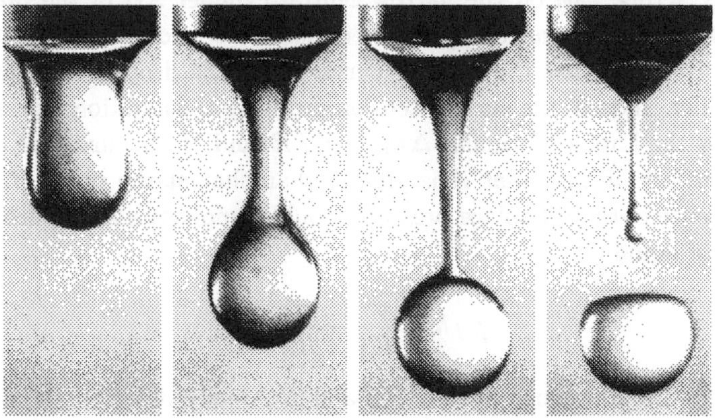

Fig. 10.8: High-speed photographs of droplet take-off.

As the droplet continues to grow gravitation becomes more prominent.

Now most of the mass accumulates in the lower part of the droplet, and the droplet begins to develop a characteristic neck (the second shot of Fig 10.8). The surface tension forces are directed vertically along the tangent to the neck, and for some time they manage to balance the droplet's weight. But not for long: at certain moment just a slight increase of the droplet's size is enough already for gravitation to overtake the surface tension and to break this balance. The droplet's neck then narrows promptly (the third shot), and off the droplet finally goes (the fourth shot). During this last stage, an additional tiny droplet forms of the neck and follows the big "maternal" one. This secondary little droplet (called the Plateau's bead) is always there, however, because of the extreme swiftness of the droplet leaving the tap, we basically never notice it.

We won't be going into details of the formation of these collateral little droplets here it is a pretty complicated physical phenomenon. We will rather try to find an explanation of the observed shape of the primary droplets in their free fall. Instant photographs of the falling droplets show clearly almost spherical shape of the little secondary drops, while the big primary ones look rather flat, something like a bun. Let's estimate the radius at which droplets start losing their spherical shape.

When a droplet is moving uniformly (at a constant speed), the force of gravity acting, say, on the narrow central cylinder AB of the droplet, Fig. 10.9, must be balanced by forces of surface tension. And this automatically requires that the radii of curvature of the droplet at A and B should differ. Indeed, the surface tension produces the additional pressure, defined by the Laplace formula: $\Delta P_L = \sigma' / R$, and, if the curvature of the droplet surface at the point A is greater than that at the point B, then the difference of these Laplace pressures could compensate the hydrostatic pressure of liquid:

$$\rho g h = \frac{\sigma'}{R_A} - \frac{\sigma'}{R_B}.$$

Let's check by how much should R_A and R_B really differ to satisfy the above relation. For tiny little droplets with radii of about $1\,\mu$ ($10^{-6}\,m$), the value of $\rho g h \approx 2 \cdot 10^{-2}\,Pa$, whereas $\Delta P_L = \sigma' / R \approx 1.6 \cdot 10^5\,Pa$! So, in this case, the hydrostatic pressure is so small when compared to the Laplacian that one can safely disregard it at all, and the resulting droplet will be very close to an ideal sphere.

But it's a completely different story for a drop of, say, $4\,mm$ radius.

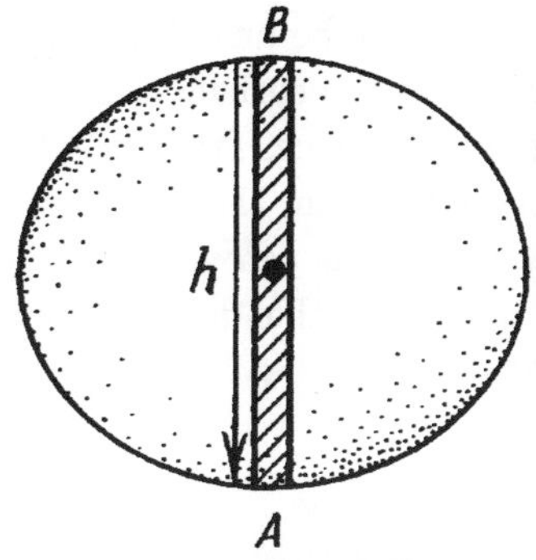

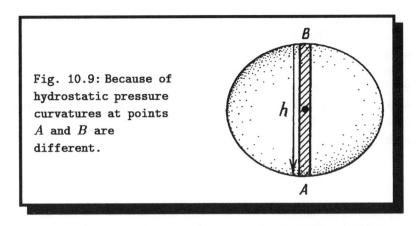

Fig. 10.9: Because of hydrostatic pressure curvatures at points A and B are different.

Then the hydrostatic pressure is already $\rho g h \approx 40\,Pa$, but the Laplacian one is $\Delta P_L = 78\,Pa$. These are the values of the same order of magnitude, and, consequently, the deviation of such a droplet from the ideal spherical shape becomes quite noticeable. Assuming $R_B = R_A + \delta R$ and $R_A + R_B = h = 4\,mm$, one finds $\delta R \sim h\left[\sqrt{\left(\frac{\Delta P_L}{\rho g h}\right)^2 + 1} - \frac{\Delta P_L}{\rho g h}\right] \approx 1mm$, and the difference of the radii of curvature at A and B now turns out to be of the same order as the size of the droplet itself.

These, just performed, calculations show us for what kind of droplets we should expect that their shape substantially deviates from the sphere. However, the predicted asymmetry turns out to be opposite to the observed in the experiment (Real droplets in photographs are flattened from the bottom!). What's the matter here? Well, the thing is that we believed the air pressure to be the same over and under the droplet. And it is really true for slow moving drops. But when the speed droplet is sufficiently high, the surrounding air does not have enough time to smoothly flow around. So there appear a region of higher pressure before the droplet and an area of lower pressure right behind it (where real turbulent vortexes are formed). The difference of the front and back pressures can actually exceed the hydrostatic pressure, and the Laplace pressure now must compensate this difference. In such circumstances, the value of $\frac{\sigma'}{R_A} - \frac{\sigma'}{R_B}$ turns negative, meaning that R_A is now greater than R_B. That (to our final satisfaction) is exactly what has been seen in the experimental pictures.

And in the end, just a short quiz about the giants and whoppers. Have you ever seen those among droplets? Not many. They simply do not survive under the normal circumstances. And for the good reason: droplets of large radii turn out to be unstable and spatter into a bunch of little ones almost instantly. It's the surface tension that assures longevity of a droplet on a hydrophobic surface. Yet once the hydrostatic pressure becomes greater than the Laplacian, the droplet spreads over the surface and breaks into smaller ones. One can use the following relation to estimate the maximal radius of a still stable droplet: $\rho g h \gg \frac{\sigma'}{R_A}$, where $h \sim R$. From which, one can find:

$$R_{max} \sim \sqrt{\frac{\sigma'}{\rho g}}.$$

For water, for instance, $R_{max} \approx 0.3\,cm$ (of course, this is just the order of magnitude estimate of the maximal size of droplets). That's why we never see, for example, really gigantic droplets on the leaves of the trees or other nonwettable surfaces.

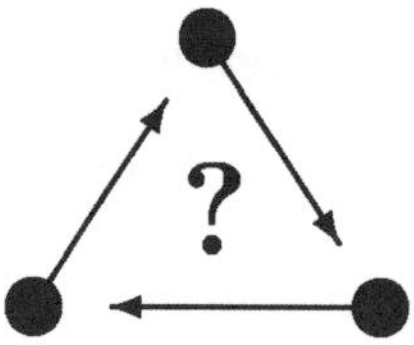

Could it happen that four soap films met at right angles along a line?

Chapter 11
The mysteries of the magic lamp

> ... "Simmetriads" appear spontaneously. Their birth resembles eruption. All of a sudden, the ocean starts coruscating as if tens of square kilometers of its surface were covered with glass. Short while later, this glassy envelope pops up and bursts outwards in a shape of a monstrous bubble, in which, distorted and refracted, arise the reflected images of the whole firmament, the sun, clouds, the horizon...
>
> *Stanislaw Lem,* Solaris.

The series of pictures, shown on the third page of the cover, haven't been taken neither on Solaris, nor from a spaceship diving in the recondite abyss of Jupiter' atmosphere, nor from the window of a bathyscaphe having dared to approach an erupting underwater volcano. Not even close. They are just photographs of working Lava-lamp, a gadget anyone could without much trouble find, for example, in a "Hands-on" toy shop or sometimes in a big department store[a]. And yet, it turns out that this seemingly simple device conceals plenty of beautiful and subtle physical phenomena.

The design of the lantern isn't very complicated. It consists of a cylinder with transparent walls, in the base of which under its glass bottom, a regular electrical bulb is mounted. The glass in the lower part is covered with a multicolor light filter, and a coiled metal wire is wrapped around the bottom perimeter Fig. 11.1. One sixth of the cylinder is filled with a

[a]It's going to cost you fifty bucks though. In the long desisted Soviet Union the same gadget would run you only one tenth of that. — It's just one more example of how much the cost of research may vary depending on where one conducts the experiments.

wax-like substance (which we will call from now on ***substance*** $\mathcal{A}$), and the rest of its volume is filled with a transparent liquid (say, ***liquid*** $\mathcal{B}$). The particular criteria for choosing these substances, as well as their properties, will be discussed a bit later, when we will be closely studying the physical processes taking place in the lamp.

It appears more convenient to conduct observations of the Lava-lantern in the dark, having it as the only source of light. So, let's turn it on and prepare to wait. As we will see, the events taking place inside the lamp could be separated into several stages. We will call the first one *"the phase of rest and accumulating the strength"*.

The substance $\mathcal{A}$ is amorphous and, therefore, does not have strict, well ordered internal structure[b]. As its temperature goes up, it becomes more and more malleable, softens and gradually turns liquid. It's worth recalling at this time a principle difference between the crystalline and amorphous substances. For the former ones, this solid-to-liquid transition (melting, in ordinary words) happens at a certain temperature point and requires a particular amount of energy (***the heat of melting***), which is expended for breaking the material's crystal structure. On the contrary, solid and liquid states of an amorphous substance aren't critically different. When the temperature is rising, amorphous materials simply soften and become liquidlike.

When turned on, the bulb of the lantern, illuminating the cylinder from the underneath through the color filter with a kind red-green glow, serves also as the heater. In the bottom floor, close to the bulb, there consequently develops a "hot spot" (the area of elevated temperature). Substance $\mathcal{A}$ in this hot region becomes softer, whereas in the same time neither the upper crust of $\mathcal{A}$, nor the liquid $\mathcal{B}$ have had enough time to warm up remaining relatively cold. As larger and larger part of $\mathcal{A}$ is softening, the top solid crust becomes thinner and thinner. Besides, due to the thermal expansion, volume of the lower, now liquid, part of $\mathcal{A}$ tends to increase, raising the pressure underneath the crust. At some moment, $\mathcal{A}$ finally breaks the crust and lunges bubbling upwards. It's like a small acting volcano is born. The quiescent phase of "rest and accumulating" is over and the new period of "volcanic activity" kicks in (see Fig. 2 on the third page of the cover).

The substances $\mathcal{A}$ and $\mathcal{B}$ are chosen in such a way that the density of the

[b]We'll dwell on the difference between crystalline and amorphous substances later in Chapter 17.

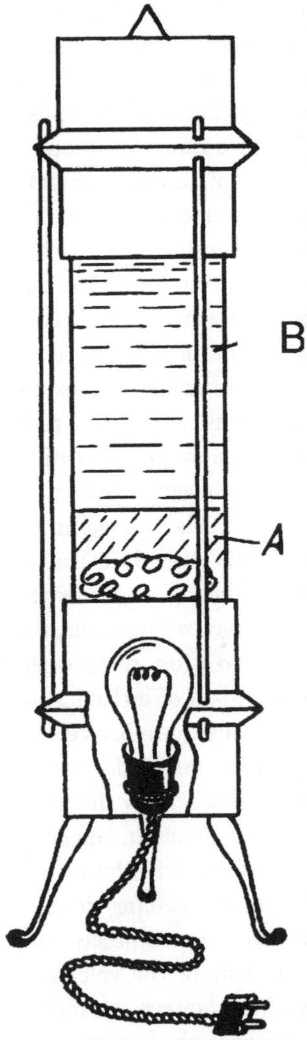

Fig. 11.1: Construction of the magic Lava-lamp.

warmed, rushing up from the crack, $\mathcal{A}$ was slightly higher than that of still rather cold $\mathcal{B}$, causing the new portions of $\mathcal{A}$, successively leaving the rift, to surface one after another[c]. On their way up, these pieces start cooling down and, when reaching surface, become solid again, assuming various and quite peculiar at times shapes. Yet now their density is back to the initial, higher than the density of $\mathcal{B}$, value, and these "smithereens" begin to sink down slowly. Some of them, usually the smaller ones, however, continue to hover by the surface for a long time. And the reason for this "recalcitrant" behavior is, of course, our old acquaintance the surface tension. Indeed, $\mathcal{A}$ and $\mathcal{B}$ are opted so that the $\mathcal{B}$-liquid does not wet the $\mathcal{A}$-solid. Hence the surface tension force acting on fragments of $\mathcal{A}$ is directed up, trying to push them out from the liquid. It's exactly the same reason why, for example, the water-striders can freely stay (and quite audaciously run) on the surface, or an oiled metal needle does not founder.

In the meanwhile, the excessive pressure in lower part of the cylinder, under the crust, has been relieved, the edges of the crack have become molten, and new portions of melted $\mathcal{A}$ are continuously trickling out of the crater. However, now they don't sever in the form of bubbles, yet rather stretch leisurely as an extended upward narrow stream. The outer surface of this stream, in contact with the cold $\mathcal{B}$, quickly cools down and stiffens, producing a sort of trunk. And if one tries to look through this trunk, one may well get surprised, for the trunk turns out to be a hollow, narrow walled tube, filled with liquid $\mathcal{B}$. The explanation would be that when the stream of melted $\mathcal{A}$ leaves the crater and runs upwards, at some point, it simply doesn't already have enough material to continue growing. Then, pressure inside the trunk decreases and, resulting from that, a crack develops somewhere by the junction between the crater edge and the trunk, and then, sure enough, the cold liquid $\mathcal{B}$ starts pouring into the cleft. The top of the $\mathcal{A}$ tube in the meantime keeps going up, and the liquid $\mathcal{B}$ fills the tube inside, cooling down and shaping the tube's inner walls, finally causing them to completely solidify.

As the vine of the volcanic plant makes it way to the surface, on the bottom of the lantern the melting continues and the next ball of the "hot" liquid $\mathcal{A}$ leaves the crater. It goes up, but now it goes up inside of the developed tube, and when it gets to the top of the tunnel, the ball, still being warm enough, extends the tube by another incremental bit. So the

[c]This resembles the famous experiment, in which a droplet of aniline, that at first peacefully rested at the bottom of a tall glass cylinder with water, was immediately starting up to the surface as soon as the temperature reached about $70^\circ\,C$ and the aniline's density became less than that of water.

plant keeps on growing, adding one by one these successive blocks (see Fig. 3 of the third cover page). Soon enough, shoving off the crusty chips of the preceding "volcanic activity", there start protruding another stem near the first one, and maybe another, after. These esoteric "underwater" plants swirl and intertwine, like those exotic shoots of jungle verdure, among the falling rocks, continuing at leisure descend from the surface; and the bottom floor, strewn with already landed boulders of $\mathcal{A}$, just adds a completing touch to this mysterious superlative happening. The picture halts for a time. We could name this stage *"the phase of the rocky forest"*.

If at this point, one turns the lantern off, the "petrified thicket" will remain there "forever" and never the lamp will be able to return to its original, two clearly segregated phases, state[d]. Yet, surprisingly though it may sound, after the kaleidoscope of the described already events, we have not still reached the working regime of our magic lamp. So let's keep it on and continue to watch.

As time goes, the liquid $\mathcal{B}$ is still warming up, the resting on the floor boulders are starting to melt again, the tangled vines of the magnificent plants are wilting down. An interesting fact: there are no really squeezed shapes among the droplets the liquefied rocks become. They all are turning out quite spherical. Under the normal conditions, the force squeezing water droplets on a hydrophobic surface is their weight. And it is balanced by the force of surface tension, tending to make the drops ideally spherical, for sphere has the minimal surface area for a given volume. In the Lava-lamp's flask, besides gravity and the surface tension, there is Archimedean buoyancy force also acting on the droplets, and, because of the closeness of the densities of $\mathcal{A}$ and $\mathcal{B}$, almost equalizing the gravitational force. So the droplets happen to be in a kind of nearly weightless situation, with nothing to preclude them from "donning" their predestined "round" habiliment (we have already discussed that topic in Chapter 10).

For a single droplet, in absence of gravity, the ideal spherical form is the most energetically favorable one. For two, or several drops, touching each other, from the same logic, it would be more beneficial to merge into one, because the surface area of a single large ball is less than the total area of surfaces of several smaller ones, of the same aggregate mass (we will let the reader to check this statement on her own). However, when looking how it works in the Lava-lantern, one notices that those almost spherical droplets

[d]Unless you switch it on again. —*A. A.*

of $\mathcal{A}$ tend to linger together without actually coupling. It seems especially striking if one remembers how promptly, almost momentarily, the mercury or water droplets, for example, couple on an unwettable surface. What does after all determine the time it takes for, say, two droplets to join?

Interestingly enough, but this question was attracting attention of different researchers and engineers since a long while ago. And not only from the point of a simple scientific curiosity. Yet also because of its critical importance for understanding physical processes in some very practical areas such as, for instance, powder metallurgy, when the preliminary powdered into grains metals are pressed and *baked* together to produce new alloys with needed physical properties. Back in 1944, the bright Russian physicist Yakov Frenkel[e] had proposed a simple yet quite useful model of such a merging process in his pioneering work, which became a fundamental in establishing the theoretical basis for this important branch of modern metallurgical technology. And now we are going to use the underlining idea of this his work to estimate the time it would take for two droplets in the magic Lava-lantern to couple.

Let's consider two identical droplets in the close proximity, so that they start touching each other. At the point of their contact, there starts, then, developing a connecting "isthmus" Fig. 11.2, which continuously grows as the two droplets merge. We will use the energy considerations (for it is the simplest and the shortest way) to estimate the coupling time. The total energy available for the system of the two drops ΔE_s results from the difference between surface energies of its initial and final states, that is the summed surface energy of the two separate droplets of radii r_0 and the big "unified" droplet of the radius r:

$$\Delta E_s = 8\pi\,\sigma\,r_0^2 - 4\pi\,\sigma\,r^2.$$

Since after merging the total volume of the drops does not change, one can write the following equality: $\frac{4\pi}{3}r^3 = 2 \cdot \frac{4\pi}{3}r_0^3$, and find $r = r_0\,\sqrt[3]{2}$, so that

$$\Delta E_s = 4\pi\,\sigma\,(2 - 2^{\frac{2}{3}})\,r_0^2. \tag{11.1}$$

[e]Ya. I. Frenkel, (1894–1952), specialist in solid-state physics, physics of liquids, nuclear physics *etc.* Probably it's worth mentioning that in 1936 Ya. Frenkel had independently of N. Bohr proposed the so-called ***drop nuclear model***. In this context coalescing droplets stood in direct relation to problems of nuclear synthesis. —*A. A.*

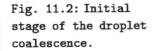

Fig. 11.2: Initial stage of the droplet coalescence.

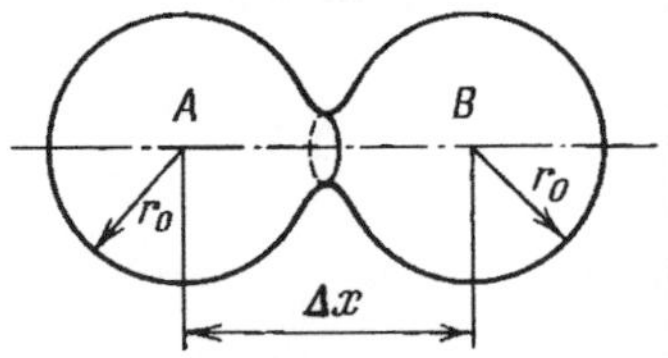

According to the Frenkel' idea, this additional energy is spent for work against forces of liquid friction, appearing in the process of redistributing droplet material as well as the surrounding medium, during the droplet merging. We can estimate this work by the order of magnitude. To find the liquid friction force, we will apply the famous Stokes' formula[f] for a spherical ball of radius R moving with velocity $\vec{v}$ in a liquid of viscosity η: $\vec{F} = -6\pi\,\eta\,R\vec{v}$. We suppose further that viscosity of the droplet material is significantly higher than that of the surrounding liquid B, which allows us to leave η_A in the Stokes' expression as the only viscosity coefficient[g]. Also, we may plug r_0 in the place of R. And then, noticing that the same quantity characterizes the scale of mutual displacement of the droplets when they merge: $\Delta x \sim r_0$. So finally, one could write for the work of the liquid friction forces:

$$\Delta A \sim 6\pi\,\eta_A\,r_0^2\,v.$$

From this expression, it's clear that the faster droplets are merging the greater amount of energy is required (because the liquid friction force increases with speed). However, the available energy resource is limited by ΔE_s, (11.1). So these two relations will provide us the searched merging time τ_F (called the *Frenkel' time* of merging). Assuming $v \sim r_0\,/\,\tau_F$ to be

[f]George Gabriel Stokes, (1819–1903), Renowned British physicist and mathematician, mostly famous for the theorem and the formula, both having commemorated his name.

[g]Sure enough the Stokes' expression was derived for a different situation, when a spherical body moved in viscous liquid. However, it's pretty obvious that in the case of two merging droplets, the liquid friction force can only depend on viscosity, droplet size and the speed of the process. Hence, from the dimension considerations, Stokes' formula turns out to be the only combination of these three physical qualities with the dimension of force (and we do not care of the exact proportionality coefficient, for our estimate is by the order of magnitude only).

the process speed, we find:

$$\Delta A \sim \frac{6\pi\,\eta_A\,r_0^3}{\tau_F} \sim 4\pi\,\sigma\,(2-2^{\frac{2}{3}})\,r_0^2,$$

and finally:

$$\tau_F \sim \frac{r_0\,\eta_A}{\sigma}.$$

For water droplets of, say, $r_0 \sim 1\,cm$, $\sigma \sim 0.1\,N/m$ and $\eta \sim 10^{-3}\,kg/(m\cdot s)$ this time turns out to be around just evanescent, $\sim 10^{-4}\,s$. Yet, for example, for the much more viscous glycerin ($\sigma_{gl} \sim 0.01\,N/m$, $\eta_{gl} \sim 1\,kg/(m\cdot s)$ at $20°\,C$) , the corresponding time is already $\sim 1\,s$, proving the fact that for different liquids, depending on their viscosity and surface tension coefficient, τ_F can vary within a rather broad range.

It's worth emphasizing at this point that even for the same liquid, due to the strong temperature dependence of viscosity, Frenkel' time can vary quite a bit. Going back to the glycerin, for instance, its viscosity drops 2.5 times when the temperature rises from 20° to $30°\,C$. The surface tension coefficient, on the other hand, stays pretty indifferent to the temperature variation, — in the considered temperature range, σ_{gl} doesn't change by more than a couple of percent. This allows us to safely assume that the temperature dependence for Frenkel's time is purely determined by the viscosity temperature dependence.

Now, let's look again at the balls of $\mathcal{A}$ lying still peacefully on the floor of the lamp through the derived estimation for the Frenkel' merging time. As long as the liquid $\mathcal{B}$ stays rather cold, $\mathcal{A}$'s viscosity remains low, and it's what keeps the balls from the "merging" rush. It is the same reason why, say, two touching wax balls do not couple into one at room temperature. Although, if one heats them hot enough, the viscosity of wax plummets and the balls merge "expeditiously". One more thing, playing an important role in the process, is the state of the surfaces of the potential partners, — the more rough and contaminated they are, the more difficult it is for the initial bridge to develop.

The merging of the $\mathcal{A}$ droplets is absolutely critical for the lantern's working cycle to go on. This explains the presence of a special means, in order to facilitate this redistribution of $\mathcal{A}$, from numerous drops into a uniform, melted mass. Sure enough, it is the mentioned already metal coil, wired along lantern's bottom perimeter. This coil is well warmed up by now, and when approaching and touching it, the droplets receive that needed heating up, lowering their viscosity and, by doing so, "warming" greatly their desire to join back the prime body of the liquefied $\mathcal{A}$. Soon enough, after all droplets finally disappear into the maternal mass, it ends up with one, unified liquid phase of $\mathcal{A}$ in the bottom of the lantern cylinder. And, because it is still continuously being heated, the liquid $\mathcal{A}$ can not remain motionless, of course. A new stage of the Lava-lamp life commences. We will call it *"the phase of protuberances"*.

Formed in the surface layers of $\mathcal{A}$, such protuberances languidly take off for their upward trek to the surface of $\mathcal{B}$, pulled, of course, by the buoyancy force, and gradually assuming, as they go, more and more spherical shape(see Fig. 5 of the third page of cover). Having reached the upper layers of $\mathcal{B}$ (where $\mathcal{B}$, due to its low thermal conductivity, still remains cold), the protuberances cool down a little, nerveless, remaining liquid this time, and begin to drown slowly, landing back on the slightly popped up surface of $\mathcal{A}$. Because of their still pretty high viscosity, it is quite difficult for them to dive into the $\mathcal{A}$ medium right away. So they bounce on its surface for while, drifting to the periphery, where the "surgical" metal coil opens up their surface and they end their live cycle exactly where they have started it.

The bulb in the cylinder base keeps heating the system, creating new and new protuberances. As temperature is continuously rising, the rate of their birth goes up as well. When taking off from the $\mathcal{A}$ surface, protuberances leave behind them smaller droplets[h], which kind of freeze perplexed in space, hesitating whether they really should go up into the unknown following their parent, or maybe just return to safety, into the original medium. In a time, a dozen of such "orphaned" liquid balls are hovering in the cylinder, some of which do finally dare to continue upwards, whereas the coy ones are descending back (Fig. 6 of the third cover page): a new stage of *"collisions and calamities"* is emerging then. And this turns out to be the longest and the most impressive phase of the lantern's activity.

The spheres are colliding, veering in various directions, however managing to avoid merging in the process. It seems like it would be advantageous, energetically, for the striking drops to couple (for the same reason we've mentioned just some paragraphs above). Yet, once again, they happen to run in the time problem. The duration of collision t, it's all they have, and

[h]By the way, these are the same Plateau balls that have been mentioned in Chapter 10.

if τ_F turns to be much longer than t, than there is no enough time for the droplets to join and they, having collided, will just simply bounce apart. Let's try to give an estimate of the collision time. Most of the collisions in the lamp are glancing ones Fig. 11.3, during which the soft liquid balls slightly deform and slide along each other. The characteristic time of such an encounter must be about $t \sim r_0 \, / \, v$. Velocity of the balls flowing in $\mathcal{B}$, v is just several centimeters per second, the ball radii are of the order of a couple centimeters too. It makes $t \sim 1\,s$, and sure enough it is too short "unite", leaving then no other option except to continue roaming aloofly and unattached in the lamp cylinder, loitering at times by the bottom, then wandering though the bulk of $\mathcal{B}$, colliding with each other, yet not merging.

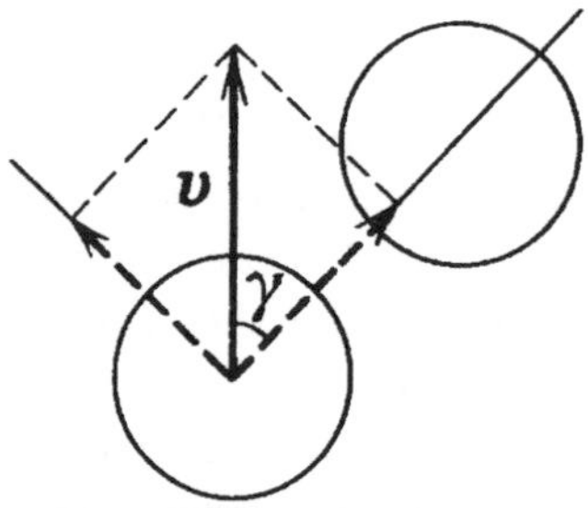

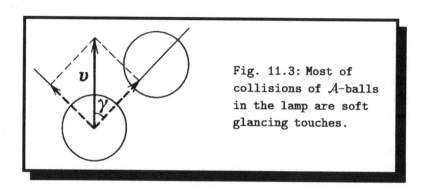

Fig. 11.3: Most of collisions of $\mathcal{A}$-balls in the lamp are soft glancing touches.

This *"phase of collisions and calamities"* can go on for hours. The manual usually recommends to switch the lantern off after 5–7 hours of operating. But under certain circumstances, when temperature of the ambient air is sufficiently high (say, you happen to marvel at the magic gadget on a sultry Austin or squashing hot Tucson summer evenings), the described "collision" stage turns out to be not the last one. Finally, after a stationary temperature distribution along the cylinder height has been attained (and the whole liquid $\mathcal{B}$ has warmed up), the densities of $\mathcal{A}$ and $\mathcal{B}$ become practically the same, and the entire $\mathcal{A}$ congregates into a single gargantuan ball. At first, the whopper is hanging in the bottom part, bouncing at times against the cylinder walls. Then, because of these "colder walls" contacts, it cools down a little, becomes a bit denser, and, consequently, sinks down to the floor. After touching the bottom though, the ball gains an extra portion of heat, its density drops again, and it returns to its previous position, where it stays until it cools down again. The cycle starts all over. This, unmentioned in the lantern's instruction, phase we could name "*the super-ball time*" (see Fig. 8 on the third page of the cover).

Finally, after we, along with our magic Lava-lamp, have gone through the numerous stages of its work, gaining some understanding of mechanism of the described processes, let's take a concluding look at these phenomena in a general, overall manner. The first question coming to mind is why do these successive, from many points repeating, events of the birth, life and death of the spheres occur at all? — It is clear, from all our previous discourse, that the driving force behind the processes is the temperature difference between the top and bottom ends of the lantern (in the thermodynamical terms, between the ***heat source*** and the ***heat sink***). If one supposes that the flux of heat propagates in the system only because of the heat conduction by the $\mathcal{B}$-liquid, the temperature of $\mathcal{B}$ will be simply changing gradually along the height, and nothing unusual, amusing would happen. The birth of spheres, on the other hand, as well as the ordinary convection, is a consequence of instabilities which sometimes develop in systems, where a thermal flow due to variation of temperature along the boundary occurs. The study of behavior and properties of these systems is the subject of rather new, though quite "volcanically" developing branch of physics called ***Synergetics***.

Chapter 12

Waiting for the tea-kettle to boil

A bright idea came into Alice's head.
"Is that the reason so many tea-things are put out here?" — she asked.
"Yes, that's it," — said the Hatter with a sigh:
"it's always tea-time, and we've no time to wash the things between whiles."

Lewis Carroll, Alice's Adventures in Wonderland.

There are thick Eastern manuscripts as well as long detailed chapters in special books devoted entirely to the tea drinking ritual. Yet when taking another, unconventional, peek at the process, one surely finds in galore interesting and edifying physical phenomena, which are not described even in the most reverent culinary "oracles".

To limber up a little, let's perform the following experiment at first. We will take two identical tea-kettles, with the equal amount of cold water (same initial temperature in both), and put them on the burners or hot plates (whatever the stove has) of the same heating power. One of the subject kettles will be covered with the lid, and the other one will stay "bareheaded". In which will the water boil first? Any housewife (no offence to the intelligence of our housewives given) will give you the correct answer right away. If she wants to have hot water faster, she will put the lid on and reply that the water will boil first in the covered kettle. Well, not taking this statement for granted, let's check it as we are supposed to, — experimentally, and wait until the water does indeed start to boil, and have discussion on the resulting observation afterwards.

In the meantime, while the two our tea-kettles are getting hot, we put one more, identical tea-kettle on a third burner. The volume of water and its initial temperature again are the same as for the two previous ones, as well as the power of the burner. Now, we aim to get water in this last tea-kettle boiling somehow faster than in the other two. How could we possibly help raising the water temperature in this kettle? A trivial way would be to stick an extra heating coil in it. Well, but let's say we do not have any available? Then let's recall that in order to increase water temperature in a vessel, it's sufficient just to add there some hotter water. Maybe it's what we need to speed up the process and reach the boiling point faster. It turns out — no. Not at all. On the contrary, it will rather slow it down. To prove this, let us assume that the original amount of water of mass m_1 at a temperature T_1, did not mix with the added water (m_2 at T_2) and did not exchange any heat with this new mass of water either. The amount of heat required for the original volume of water in the kettle to boil would be $Q_1 = cm_1 (T_b - T_1)$ where c is the specific heat of water. But now, besides this energy, the mass m_2 has to be heated to the same boiling temperature, so that the total amount of heat needed is:

$$Q_1 = cm_1 (T_b - T_1) + cm_2 (T_b - T_2).$$

Even if we pour boiling water straight into the tea-kettle, still the added water will cool down somewhat during the "trans-filling", lowering its temperature below T_b. Obviously our quite naive conjecture that the two portions of water in the same pot remain unmixed after the addition did not affect in any way the law of conservation of energy in the system, yet permitted us to handle the evaluation simpler and faster.

By the time we've finally given up the idea of adding water into the third kettle, the first of our two "original" tea-kettles begins hissing. And what would be the physical mechanism behind this so familiar sibilant sound, and what is its characteristic frequency? Next, we shall try to answer these questions.

As the first candidate to generate this whistling discord, one could suggest oscillations excited in the liquid when the "ripened" steam bubbles take off from the walls and bottom of the reservoir (tea-kettle in this case). These bubbles always start developing on all sorts of microcracks and other defects, ever present on any real surface. The typical size of such bubbles, before the water starts to boil, is about $1\,mm$ (after that they can reach as much as $1\,cm$). To estimate the frequency of sound produced by simmering liquid, we need to know how long it takes for the bubbles to release from the bottom. This time actually measures the length of the push the liquid experiences when every next bubble takes off, and, therefore, the period of vibrations they excite. Within our assumptions, the searched frequency is determined by the reciprocal of this time: $\nu \sim \tau^{-1}$.

When the nascent bubbles are resting on the bottom, there are two forces acting upon them[a]: the Archimedean buoyancy force, $F_A = \rho_w g V_b$ (here V_b is the volume of the bubble and ρ_w is the density of water) pushes it upward and the surface tension that keeps it attached to the surface, $F_S = \sigma l$ (where l is the border length of the contact area between the bubble and the surface). As the bubble grows (V_b increases), the Archimedean force is rising too, and, at a certain moment, it exceeds the retaining force of surface tension. The bubble takes off, starting on its journey up, Fig. 12.1. Hence, the resulting force acting on the bubble during its "departure" stage, should be of the order of F_A. Whereas, the acceleration of a bubble in liquid is defined, of course, not by its own negligible mass (consisting mostly of the mass of air trapped in it), but by the mass of liquid involved in the motion. For a spherical bubble this so-called ***associated mass***, equals $m^* = \frac{2}{3}\pi \rho_w r_0^3 = \frac{1}{2}\rho_w V_b$ (r_0 being the bubble's radius).

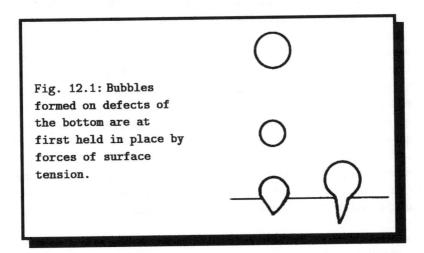

Fig. 12.1: Bubbles formed on defects of the bottom are at first held in place by forces of surface tension.

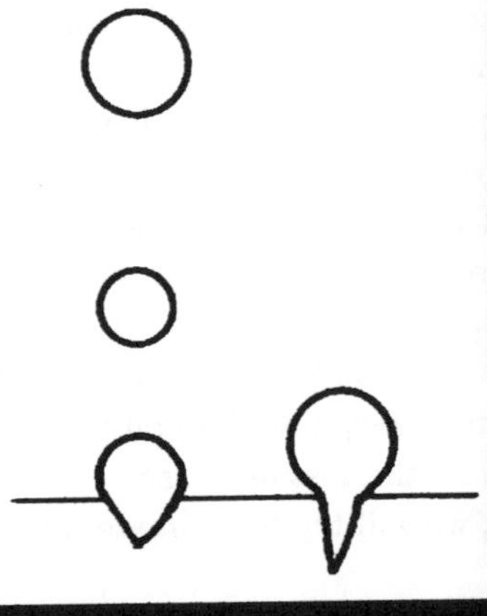

Thus, for the acceleration of the bubble during the initial stage, one

[a]Here we disregard the tiny weight of the bubble.

finds:

$$a \sim \frac{F_A}{m^*} = 2g.$$

Now we can evaluate the bubble release time, considering (again, for simplicity sake) the motion to be uniformly accelerated. Our subject bubble will climb up to the height comparable to, say, its size within the time of

$$\tau_1 \sim \sqrt{\frac{2r_0}{a}} \sim 10^{-2}\, s. \tag{12.1}$$

Then, the corresponding characteristic frequency of the generated at the bubble's take off sound should be equal to $\nu_1 \sim \tau_1^{-1} \sim 100\, Hz$. This seems like maybe an order of magnitude less[b] than the sibilant tune one hears when a tea-kettle is being heated on a stove (long before though the water starts actually boiling)[c].

So it turns out that there must be another cause for the tea-kettle's hissing, when it's warming up. To find out this second reason, one would have to closely follow the bubble's fate after it leaves its parental surface. Having taken off from the hot wall (or bottom) where the vapor pressure in the bubble was around atmospheric (otherwise it couldn't expand enough to start making it upwards), our protagonist hurries up in the higher, and, naturally , still colder, layers of water. So, the saturated water vapor, filling the bubble, cools down too, causing the inside bubble pressure to drop and not be able any more to compensate the external pressure of liquid exerted on the poor bubble. As the result, the squashed bubble flops or gets squeezed into a tiny one (the latter happens if, besides water vapor, it had also a little bit of air inside), generating a sound pulse in the liquid, Fig. 12.2. This very process of massive "death or bad maiming" of numerous steam bubbles on their way to the water surface is indeed sensed by us as the hissing noise . And now, following our already established habit, we will try to estimate its frequency, of course.

[b]Frequencies in the $100\, Hz$ range are familiar from the humming noise of old grandfather's radios.—*A. A.*

[c]Note that the surface tension did not make it into the (12.1) indicating in a way that the bubbles produce sound not only when leaving the surface but during all their accelerated motion upwards. This lasts until the buoyancy force gets compensated by the proportional to velocity viscous friction.

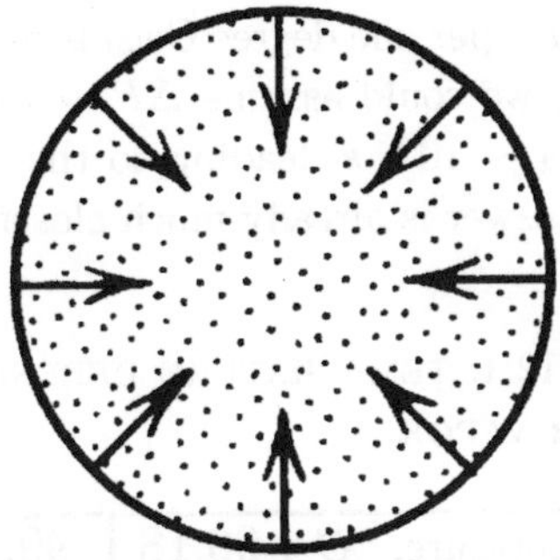

Fig. 12.2: Before the active boiling starts collapses of hundreds of tiny bubbles make the tea-kettle sing.

From the Newton's second law equation, for the mass m of water rushing into the bubble when it collapses, we can write:

$$m\,a_r = F_p = S\,\Delta P.$$

Here $S = 4\pi\,r^2$ is the area of the bubble's surface, F_p is the total pressure squeezing the bubble, ΔP is the pressure difference across the bubble's envelope and a_r is the inward acceleration of the envelope. It's pretty clear that the mass involved in the process of "squeezing" should be of the same order as the product of water density times the bubble's volume: $m \sim \rho_w\,r^3$. So, we could rearrange the Newton's equation in the following manner:

$$\rho_w\,r^3\,a_r \sim r^2\,\Delta P.$$

Further, neglecting the fraction of pressure arising from the bubble's surface curvature, as well as the smidgen of air possibly ensnared inside the bubble, we will consider ΔP to be constant (more precisely, depending only on the temperature difference between the bottom and surface layers of water in our tea-kettle). Now, evaluating the acceleration ar as $a_r = r'' \sim r_0\,/\,\tau_2^2$ where τ_2 is the "flop" time we are looking to for, we find that

$$\rho_w \frac{r_0^2}{\tau_2^2} \sim \Delta P,$$

which gives

$$\tau_2 \sim r_0\sqrt{\frac{\rho_w}{\Delta P}}. \qquad (12.2)$$

Near $T_b = 100° C$, the saturated water vapor pressure drops by about $3 \cdot 10^3\, Pa$ per one degree Celsius of the falling temperature (see Table 12.1). Hence, we could assume $\Delta P \sim 10^3\, Pa$, and finally write for the searched time $\tau_2 \sim 10^{-3}\, s$, leading to the noise frequency of $\nu_2 \sim \tau_2^{-1} \sim 10^3\, Hz$. This answer is already much closer to the value perceived by our ears[d].

Table 12.1: TEMPERATURE DEPENDENCE OF PRESSURE OF SATURATED WATER VAPOR.

Temperature, $°C$	96.18	99.1	99.6	99.9	100	101	110.8
Pressure, kPa	88.26	98.07	100	101	101.3	105	147

One more fact supporting our conclusion that the proposed mechanism is indeed responsible for the "tea-kettle" noise is that, according to (12.2), its characteristic (high) frequency goes down with the temperature growing. Right before the boiling begins, the bubbles cease flopping even in the upper layers of water. Then the only remaining sound becomes that produced by the bubbles taking off from the bottom. The frequency of the "tune" drops noticeably when water in the kettle is about to start boiling. After it finally happens, the kettle's "voice" may change again though, especially if one opens the lid: the gurgling sound we hear is generated now by the bubbles rupturing right by the water surface. This pitch depends also on how the water level, as well as on the kettle's shape[e].

Thus, we have established that the kettle's noise before water in it starts boiling is related to hundreds of steam bubbles, produced on the hot bottom, departing to the surface, and then perishing in the upper, still not hot enough, layers of water. All these processes become especially visual if one is heating water in a glass pot with transparent walls. Let's not hurry though to congratulate ourselves that we were the first to sort things out about this interesting question of the "singing" boiling water. Long while before us, back in the eighteenth century, the Scottish physicist

[d]According to Fig. 12.3 the temperature drop (and the corresponding pressure difference ΔP and frequency ν) may be up to the order of magnitude higher. —*A. A.*

[e]One more argument in favor of the proposed mechanism is that bubbles in fizzy drinks don't make sound we have been talking about. The difference with the boiling water is that the carbon dioxide filled bubbles can not collapse. —*A. A.*

Joseph Black[f] was studying the phenomenon and had established that the sound was produced by a duet consisted of the ascending to the surface steam bubbles and the vibration of the vessel's walls.

Now comes the time for water in the first of our subject tea-kettles (the one covered with lid, remember?), the predicted favorite winner, to boil. The moment is quite "resolutely" announced by the stream of steam jetting out of the kettle's spout. And what could be the velocity of the stream, by the way? One can solve this, honestly not the most challenging, problem, noticing that during the steady boil, all the energy provided to the kettle, is being spent for vaporizing of water. Let's think that in this case the only way produced steam is escaping outside is through the spout. Suppose further that a mass ΔM of water is vaporized in time Δt, on the expense of the supplied to the system heat. Then, one could write the following balance equation:

$$r\,\Delta M = \mathcal{P}\,\Delta t,$$

where r is the specific heat of vaporization (heat per unit mass), and $\mathcal{P}$ is the power of the heater. During the same time Δt, the same mass ΔM is supposed to leave the kettle through its spout, otherwise produced vapor would accumulate under the lid. If the area of the outlet orifice (pretty much the perpendicular cross-section of the spout) equals s, the steam density is $\rho_s(T_b)$, and v is the velocity in question, then the following relation holds:

$$\Delta M = \rho_s(T_b)\,s\,v\,\Delta t.$$

From the above table, we pick $\rho_s(T_b) = 0.6\,kg/m^3$ as the density of saturated water vapor at $T_b = 373^\circ\,C$. If one doesn't happen to have a suitable table handy, one can use for this purpose the Clapeyron-Mendeleev's gas

[f]J. Black, (1728–1799), Scottish physicist and chemist; the first to point out the difference between the heat and temperature; introduced the idea of heat capacity.

law[g]:

$$\rho_s(T_b) = \frac{P_s(T_b)\,\mu_{\mathrm{H_2O}}}{RT_b} \approx 0.6\,kg/m^3. \tag{12.3}$$

Thus, for the velocity of steam escaping the spout, one finds:

$$v = \frac{\mathcal{P}RT_b}{r\,P_s(T_b)\,\mu_{\mathrm{H_2O}}\,s}.$$

After the required substitutions: $\mathcal{P} = 500W$, $s = 2\,cm^2$, $r = 2.26\cdot10^5\,J/kg$, $P_s(T_b) = 10^6\,Pa$ and $R = 8.31\,J/1^\circ\cdot mole$, the searched velocity is $v \sim 1\,m/s$.

Now, finally, water goes on boiling in the second (open) tea-kettle. It has noticeably lagged behind the "capped victor". One should be very careful taking this one off from the stove, — if you just grab it by the handle, you can very easily scald yourself (we are sure, though, that our reader does understand that safety is first in any kind of experiment). Anyway, our next question is the safety (rather "unsafety") related one: what does scald worse — the steam or the boiling water? Adding to this question some required physical strictness, we could paraphrase it in the following, way: what does scald more badly - a certain mass of steam or the identical mass of boiling water. Well, for the answer we will have to do some estimations again.

Suppose there was $V_1 = 1\,l$ of saturated, one hundred degree (Celsius, of course), steam under the lid of the tea-kettle. Imagine further that after the lid is opened, one tenth of this steam condenses on the "unlucky" hand. We know already that the density of water vapor at $T_b = 100^\circ C$ equals $0.6\,kg/m^3$. Hence, the mass of the condensed on the hand water is going to be $m_s \approx 0.06\,g$. The amount of heat produced during the condensation and subsequent cooling of the condensed water from $100^\circ C$ down to the room temperature $T_0 \approx 20^\circ C$, then, will be $\Delta Q = r\,m_s + c\,m_s(T_b - T_0)$. And, therefore, as $c = 4.19\,J/kg$ and $r \approx c\cdot 540^\circ C$, it takes about ten times more of the boiling water than the hot steam to get the same thermal

[g]The expression for gas density ρ can be obtained from the equation of state of the ideal gas $PV = \frac{m}{\mu}RT$, where P, T, and m are the pressure, temperature and mass of gas enclosed in the volume V respectively. This can be easily brought to the form

$$\rho = \frac{m}{V} = \frac{P\mu}{RT},$$

where μ is the molecular mass of the gas and R is the gas constant per mole.

effect! Besides, the affected by the scalding steam area (due to much higher mobility of the vapor molecules) is always significantly larger than that if a hot water is poured on a surface. Thus, the answer to our question is that the steam would be unequivocally much more dangerous as the burning agent than the water of the same (boiling) temperature.

However, performing all these dangerous evaluations, we have swerved quite a bit from our original test with the two tea-kettles. Why did it take water in the opened one so much longer to start boiling after all? Let's look at the phenomenon closer. The answer seems to be almost trivial: during the heating , the nimblest molecules of water (those with higher velocity) can, rather freely, escape from the open kettle, pilfering some energy from the water remaining inside and, by doing so, effectively cooling it (this process is nothing but evaporation). Hence the heater, in this case, is supposed to provide not only energy to warm water to the boiling temperature, yet also the heat required to evaporate some of the water. Thus, it's clear that it is going to take more energy (and, therefore, time, because the heater power is fixed) than just heating the water in the capped kettle, where those swift fugitives have no other options but to congregate under the lid, building up the saturated water vapor, and eventually go back to water, turning in the "stolen" energy surplus.

Yet, there are two more effects occurring at the same time, although opposite to the above one. First, during the evaporation, the mass of water which needs to be warmed to T_b somewhat drops; secondly, in the open vessel, the pressure over water is atmospheric and, hence, the boiling process starts at exactly $100° C$. On the other hand, in the covered kettle, if it's filled sp that the steam can't make it out through the spout, the pressure above the water surface, because of the intensive vaporizing, will rise. Note that now it's the sum of the partial pressure of a small amount of air under the lid as well as of the steam itself. With increasing external pressure the boiling temperature should rise too, for it is determined by the equality of the saturated vapor pressure inside the developing in liquid bubbles to the external pressure. So, which of these factors should we prefer, choosing as the decisive one?

Each time when such an uncertainty comes, one should retreat to a precise calculation or, at least, evaluate the magnitudes of the involved effects. So, at first, let's estimate the amount of water which escapes from the open tea-kettle before it starts boiling.

Molecules in liquids strongly interact with each other. Yet, if in crystals the potential energy of molecules is much greater than their kinetic energy and in gases, on the contrary, the kinetic part dominates, in liquid the potential and kinetic energies are of the same order. So molecules in the liquid most of the time fluctuate around some "ascribed" to them equilibrium positions, yet once in a while managing to hop to a different neighboring equilibrium location. "Once in a while" just means far longer duration compared to the period of oscillations about equilibrium points. However, in our conventional time scale, such hops recur indeed quite frequently: in just the jiffy of one second, the "jittery" molecule of liquid can change its equilibrium position billions of times!

However not every single molecule which happens to be wandering by the surface of the liquid, can actually escape from it. To finally free themselves, such molecules ought to spend some energy performing a certain work against the interaction forces. One may say that the potential energy of a molecule of water is less than that of a molecule of steam by the amount of heat of evaporation, normalized per single molecule. Then, if r is the specific heat of evaporation, the molar evaporation heat is μr , and the "molecular" heat of evaporation will be $U_0 = \mu r \,/\, N_A$, N_A being the Avogadro number. This work is performed on the expense of the kinetic energy of molecule's thermal motion E_k. The corresponding average value $\bar{E}_k \approx kT$ ($k = 1.38 \cdot 10^{-23}\, J \,/\, 1^\circ K$ is the Boltzmann[h] constant) turns out to be far less than U_0. Nevertheless, according to the laws of molecular physics, there is always some number of molecules with kinetic energies high enough to overcome the attraction forces and flee away. The concentration of these extremely swift molecules is given by the following expression:

$$n_{E_k > U_0} = n_0\, e^{-\frac{U_0}{kT}}, \tag{12.4}$$

where n_0 is the total density of molecules, and $e = 2.7182\ldots$ is the base of natural logarithms.

Now for a time being, we shall forget about the hops of molecules in liquid, and consider these high energy molecules as a gas. A molecule of such a gas can reach the surface from inside in a short instant Δt provided its velocity v is directed outwards and it has started less than from $v\,\Delta t$ away from the surface. For a surface area S these are the molecules from

[h]L. Boltzmann, (1844–1906), Austrian physicist, one of the founders of classical statistical physics

the cylinder of the height $v\,\Delta t$ with the base S. Let us assume for simplicity that $\sim \frac{1}{6}$ of all the molecules in the cylinder, that is $\Delta N \sim \frac{1}{6} n\, S\, v\, \Delta t$, move towards the surface. Taking the density of molecules with energies greater than U_0 from equation (12.4) we obtain for the evaporation rate (that is the number of molecules that escape the liquid in unit time):

$$\frac{\Delta N}{\Delta t} \sim \frac{n\, S\, v\, \Delta t}{6\, \Delta t} \sim S\, n_0 \sqrt{\frac{U_0}{m_0}}\, e^{-\frac{U_0}{kT}},$$

where we have taken the velocity $v \sim \sqrt{U_0\,/\,m_0}$. So the mass carried away from the liquid per unit time is equal to:

$$\frac{\Delta m}{\Delta t} \sim m_0\, \frac{\Delta N}{\Delta t} \sim m_0\, S\, n_0 \sqrt{\frac{U_0}{m_0}}\, e^{-\frac{U_0}{kT}}. \tag{12.5}$$

It turns more useful to recalculate this mass normalizing it to the $1^\circ\, K$ temperature increase, while the kettle is being heated. To do so, we will use the energy conservation law: our kettle, in a time Δt, receives from the burner the amount of heat $\Delta Q = \mathcal{P}\, \Delta t$ ($\mathcal{P}$ is the burner's power), and , consequently, the water temperature rises by ΔT, entailing:

$$\mathcal{P}\, \Delta t = c\, M\, \Delta t,$$

where M is the mass of water in the kettle (here we disregard the heat capacity of the kettle itself). After plugging $\Delta t = c\, M\, \Delta T\, /\, \mathcal{P}$ into the equation for evaporation rate (12.5) we have:

$$\frac{\Delta m}{\Delta T} = \frac{\rho\, c\, S\, M}{\mathcal{P}} \sqrt{\frac{r\, \mu_{H_2O}}{N_A\, m_0}}\, e^{-\frac{r\, \mu_{H_2O}}{N_A\, k\, T}} = \frac{\rho\, c\, S\, M\, \sqrt{r}}{\mathcal{P}}\, e^{-\frac{r\, \mu_{H_2O}}{N_A\, k\, T}}.$$

As the tea-kettle is being heated, the temperature scans from the room to the boiling, $373^\circ\, K$, value. However, we conjecture at this point (and rightfully) that the majority of the mass is being lost while the temperature of water is already pretty high, close to its boiling value, so that we could put, say, $\bar{T} = 350^\circ\, K$ as the average temperature into our exponential expression above. For the rest of the terms we assume: $\Delta T = T_b - T_0 = 80^\circ\, K$, $S \sim 10^{-3}\, m^2$, $\rho \sim 10^3\, kg/m^3$, $\mu_{H_2O} = 0.018\, kg/mol$, and $c = 4.19 \cdot 10^3\, J/kg$. After putting all these values into the formula, we finally find:

$$\frac{\Delta m}{M} \approx \frac{\rho\, c\, S}{\mathcal{P}}\, \sqrt{r}\, e^{-\frac{r\, \mu_{H_2O}}{R\, \bar{T}}}\, (T_b - T_0) \approx 3\,\%.$$

Thus, while being warmed up to the boiling temperature, just a few percent of the total mass of water leaves actually the tea-kettle. The evaporation of such a mass takes an extra energy from the heater, and, naturally, protracts the heating before it reaches the boiling point. To understand by how much, one could calculate and find out that the vaporization of this amount of water requires from the heater the amount of energy equivalent to the heating from room to boiling temperature of about one forth of the total mass of water in the tea-kettle.

Now let's turn to the second (or maybe the first — we don't remember already), covered with the lid, kettle, and look closer at the effects slowing its reaching the boiling condition. The first of them (potential change of mass of water during the heating) should be dropped off right away, for, as we have just showed, the evaporation of about 3 % of water is energetically equivalent to the heating of about 25 % of water and, therefore, the heat required to bring these extra 3 % of water mass to the boil in the closed tea-kettle, could be disregarded.

It turns out that the second of the effects (the increase of pressure over the water in the covered vessel) can not play any noticeable competitive role either. Indeed, if the tea-kettle is completely filled with water (steam can't escape from the spout), then, the additional (to the atmospheric) pressure obviously can not exceed the lid's weight divided by its area, for , otherwise, the lid would start to jounce releasing the steam. Assuming $m_{lid} = 0.3\,kg$ and $S_{lid} \sim 10^2\,cm^2$, we can limit this extra pressure by:

$$\Delta P \leq \frac{m_{lid}\,g}{S_{lid}} \approx \frac{3\,N}{10^{-2}\,m^2} = 3 \cdot 10^2\,Pa.$$

And having checked once more in Table 12.1, one finds that such an increase shifts the boiling temperature by not more than just $\delta T_b \sim 0.5^\circ\,C$. Hence, to make the water boil, it would take an extra heating energy of $\delta Q = c\,M\,\delta T_b$. Comparing this to $r\,\Delta M$, we see that the inequality $r\,\Delta M \gg c\,M\,\delta T_b$ holds with the safe ratio of at least as much as 30:1. The fact allowing us to conclude that the increase of boiling temperature of water in the fully filled tea-kettle covered with the lid can not seriously compete (regarded energywise) with the evaporation of water from the open water surface in the "bareheaded" vessel.

Digressing a bit, we would like to mention here that the just described phenomenon of increasing pressure during the heating of liquid in a closed volume has been successfully employed in the design of a utensil, called the "pressure cooker" (familiar maybe to those who still do real cooking, at least once in a while). Instead of the spout though, it has a safety relief valve, which opens only if the pressure inside goes over a certain limit, the rest of the time, such a vessel remains entirely sealed. As liquid inside gets vaporized and all the steam is kept in the cooker, the internal pressure rises to about $1.4 \cdot 10^5\,Pa$, before the relief valve opens up, so that the boiling temperature (going back to our useful Table 12.1) moves up to $T_b^* = 108°\,C$. This allows to cook food much faster than in a regular pot. However, one should be extremely cautious while opening the pressure cooker after taking it off the stove, because when unsealed, the inside pressure drops and the liquid becomes significantly ***superheated***. Therefore a certain mass δm, such that $r\,\delta m = c\,M\,(T_b^* - T_b)$, will instantly evaporate, and can cause a bad scalding. (In this situation, the liquid starts boiling explosively in the whole volume of the pot at once).

By the way, at high elevations (additionally to the typical beautiful scenery), the atmospheric pressure is lower and it becomes quite an undertaking to cook, say, a piece of meat if water starts already boiling at $70°\,C$ (the ambient pressure at the elevation of Everest, for example, is about $3.5 \cdot 10^4\,Pa$). So the pressure cooker is usually a quite welcomed part of the climber's equipment. Because of the ability to reach an acceptable cooking temperatures, it saves the fuel as well. That is another weighty justification to have this massive utility in the backpack.

But let's go back to our kettles, boiling now at full tilt, still on the stove. It's time to take them off. Take a notice that the one with the lid doesn't stop boiling right away, after having been taken from the stove: the steam continues to puff out for some time. What fraction of water does actually evaporate during this ("postheating") boil?

To answer this one, we have to look at the chart in Fig. 12.3, representing the dependence of temperature of water on height while water is boiling (the heat, of course, is being supplied through the bottom). From the picture we see that the slim bottom layer of about $\Delta H = 0.5\,cm$ thick is quite hot, with the temperature drop across it from $T_{bot} = 110°\,C$ (T_{bot} is the bottom temperature) to $T_i = 100.5°\,C$. The rest of the water stays, according to the graph, at about $100.5°\,C$, yet undergoes a next step down of $\Delta T = 0.4°\,C$, when approaching the free surface of the liquid (the graph corresponds to the water level in the kettle being $H = 10\,cm$). Thus, for the amount of heat (extra to the equilibrium) stored in the vessel after its heating has been stopped, we can write the following formula:

$$\Delta Q = c\rho S\,\Delta H\left(\frac{T_{bot}-T_i}{2}\right) + c\rho S\,H\,\Delta T,$$

where S is the kettle's bottom area (the kettle is considered to have a cylindrical shape). Sure enough, the tea-kettle's base is overheated somehow also, but because of the much higher specific heat of water, we can safely disregard the contribution of this effect.

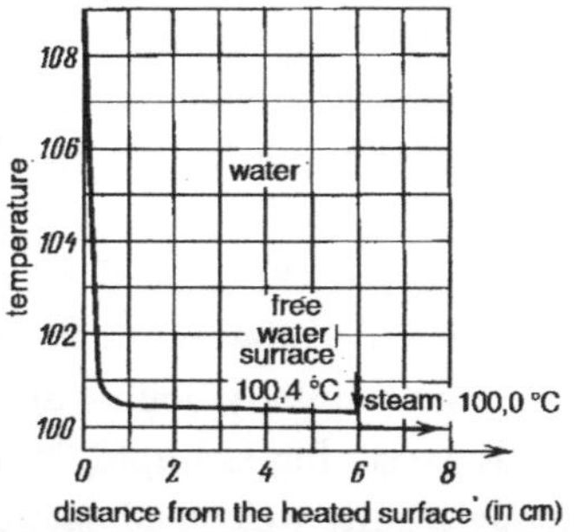

Fig. 12.3: Temperature near the bottom of boiling kettle is much higher than that of the mass of water.

The ΔQ amount of extra heat is then spent for evaporating the layer of liquid of thickness δH. The mass δm of such a "slice" of water can be found from the equation of heat balance:

$$r\,\delta m = \rho S\,\delta H\,r = c\rho S\left[\Delta H\left(\frac{T_{bot}-T_i}{2}\right) + H\,\Delta T\right]$$

consequently, giving us for δH:

$$\frac{\delta H}{H} = \frac{c}{r}\left[\frac{\Delta H}{H}\left(\frac{T_{bot}-T_i}{2}\right) + \Delta T\right] \approx 2\cdot 10^{-3}.$$

So it shows that after we take the kettle from the stove, it still looses, due to continuing boiling, about 0.2 % of its water content.

The typical time it would take to boil out all water of, say, mass $M = 1\,kg$ from a kettle provided with heat of $\mathcal{P} = 500\,W$, can be calculated as:

$$\tau = \frac{r\,M}{\mathcal{P}} \approx 5\cdot 10^3\,s.$$

Therefore, the 0.2 % of its mass will be vaporized within about 10 s (supposing that the evaporation rate does not change with respect to the stationary regime).

And now, after all this discussion, there comes at last the time to serve our a bit too long awaited tea. By the way, in the Eastern countries, it's customary to drink tea using tiny tea bowls rather than tea cups or tea glasses. The former were first introduced most likely by the nomadic tribes of Asia — the small bowls are much easier to pack and they are far less fragile, which are obvious conveniences when one has an "itinerary" lifestyle. Besides, they have one more serious advantage over the ordinary glasses: the bowl shape, wider at the top, lets water cool down faster in the upper layers precluding a possible burning by the "scorching" liquid, while the tea in the lower part of the utensil remains still hot.

However in Azerbaijan you may meet another type of the tea-drinking vessel, called Armudi Fig. 12.4. Here the widening top part helps the safe and pleasant cooling of the beverage whereas the spherical part below has the minimum surface and therefore keeps the tea hot for longer time. Thus you may enjoy sipping the full-flavored warm tea as the long oriental table-talk goes on.

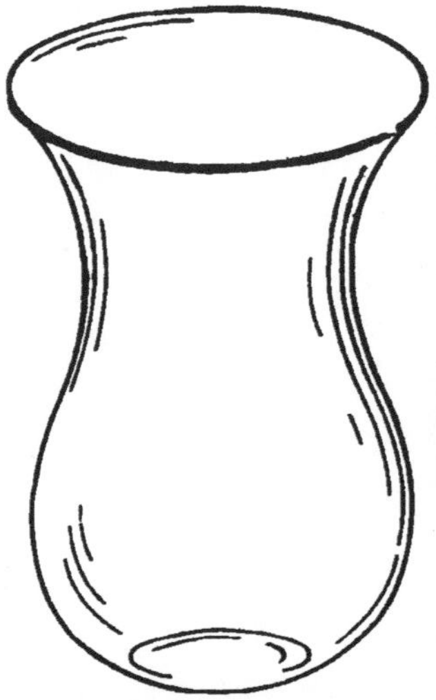

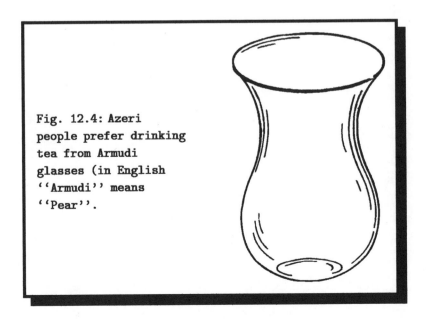

Fig. 12.4: Azeri people prefer drinking tea from Armudi glasses (in English ''Armudi'' means ''Pear''.

The nifty porcelain tea-cups (not those whopper-mugs you are usually getting as safety awards from your company), which have been around since the centuries ago, most of the time have that widening at the top profile too. The less advanced from this point of view cylindrical glasses came into general use as the tea serving ware only in the nineteenth century, simply because of their lower cost; and traditionally were used by men, whereas the artful china tea cups were politely left for the best half of the human race. Over time though, this inferiority of the tea-glasses was fixed a little bit with the invention of glass-holders (often back then decorated with their owner's monograms).

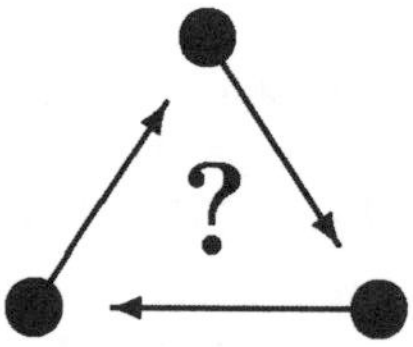

Could you think of what physical criteria material for glass-holders should satisfy? Would, say, aluminum and silver be the good candidates?

Chapter 13
Craving microwaved mammoth

I could not allow such a wonderful wildfowl to escape from me and I loaded my gun with an ordinary ramrod. Then I shot it straight through all of the partridges as they always rise from the ground in a direct line before each other. The rod had been made so hot with the shot that the birds were completely roasted by the time I picked them up.

E. Raspe, The adventures of Baron Munchausen.

The day when the Neanderthal man tamed the fire opens *era humana.* Once and forever the MANKIND broke off with the ape ancestors. Fire opened the way to smelting metals, manufacturing cars, flying to the outer space but above all... it allowed to forget the taste of raw meat. A well-browned beefsteak may symbolize civilization side by side with the model of atom.

The civilization grew older and methods of food preparation changed. The camp-fire under a spitted mammoth[a] was succeeded by the hearth, then came the time of wood-, coal- and gas-stoves, those were replaced by primuses, hotplates, electric stoves, grills, toasters, roasters...

The interaction of fire with food changed in appearance but the physical entity of the process stayed almost untouched: the heat required for cooking was either transferred immediately (sometimes with the help of convection) or by means of infrared radiation. The first mechanism works, for example,

[a]Think whether it is really possible to grill a whole mammoth? Read about the preparation of elephant legs by african tribes in the book "The plant hunters" by Mayne Reid.

when making diet cutlets or "manty"[b] in the *steamer.* where food is cooked by hot steam rising from the water that boils below. Cooking soup uses two effects at once: direct heating at the bottom of the pot is followed by convective mixing of upper and lower layers of the liquid. In the mean time electric grill or charcoal barbecue are the examples of radiative heat transport.

The evolution of the "fireplace" (let this denote the heat source) notably affected kitchen technology and cooking recipes. Possibilities of making new exquisite and delicious dishes arose. But justice calls us to confess that along with the enhancement of culinary arsenal .the development of the fireplace drove many dishes away from the table of the mankind. Some of those were forgotten, others yielded to the ersatz. Say, the true *Pizza Napoletana* can be baked in a few minutes almost of nothing but **only** in a special blazing wood-stove. Therefore old *pizzerie* are proud not only of their long father-to-son tradition but of the *forno* (this means "stove" in Italian). There you may watch the rite of pizza-making from the very take-off. *Pizzaiolo* dexterously sculptures your pizza, sets it with a wooden shovel into the stove, a moment, and here it comes, covered with sizzling hot cheese, mouth-wetting, calling to be immediately devoured with the best beer[c]. But watch out! If hunger catches you in front of some ultra-modern PIZ-Z-ZA place with a luxurious interior glittering of mirrors but lacking... Yes, right you are, lacking good old wood-stove, please, stand the temptation. Let nobody decoy you, it's safer to miss the lunch and escape a pretentious imitation served there.

However we got too far from the initial subject. Let us return even if not right away to the kitchen then... to the lecture on history of metallurgy. There you will learn that furnaces at ironworks were changing almost as often as culinary appliances. In particular the Faraday's[d] discovery of the law of electromagnetic induction opened doors to invention of induction melting. Here is a sketch of the idea: a piece of metal is placed into a strong rapidly changing magnetic field. As metals are good conductors the

[b]"Manty" is the Uzbek speciality made of minced seasoned lamb wrapped in thin dough like Italian tortellini.

[c]In Naples pizza is indisputably served with beer.

[d]M. Faraday, (1791–1867), English physicist and chemist.

Fig. 13.1: Alfonso, the famous pizzaiolo of Naples (right), gives his friend Andrei Varlamov (left) a lesson in front of his *forno*.

inductive electromotive force (EMF) that appears due to the variation,

$$\mathcal{E} = -\frac{d\Phi}{dt}, \tag{13.1}$$

(Φ is the magnetic flux penetrating the specimen) gives rise to ***eddy induction currents***. or, otherwise, ***Fucault currents***. The word "eddy" indicates that the currents are closed because they follow closed lines of the induction electric field. Induction currents generate the Joule[e] heat like any other currents caused by applied electric field. If the EMF of induction is big enough (this requires big amplitude and high frequency of the magnetic field) the evolved heat will suffice for melting the metal. Induction melting is widely used in production of high-alloy steel, aerospace metallurgy *etc.*

Well, but even astronauts get hungry... Let us leave the orbital vacuum induction furnace and visit the kitchen module of the spaceship. Here we shall find a fireplace of the sort that drops out of the long sequence listed

[e]J. P. Joule,(1818-89), English physicist, specialist in thermo- and electrodynamics.

above. This oven rather resembles an induction melting facility than a conventional kitchen utensil. It heats the food with the help of ultrahigh frequency electromagnetic radiation.

It was already in the sixties that astronauts who spent more and more time in orbit got fed up with tubed food. However, for conspicuous reasons, taking primus to space was absolutely out of question. First, the flame would consume the priceless oxigen and, second, the ruthless weightlessness would have a frustrating effect on the earthly magic of old appetizing recipes. (Just try to imagine the cooking of weightless soup on primus. What other difficulties would you predict to space cooking?)

The way out was found in using a kitchen analogue of induction furnace. Remember that almost all human food contains a noticeable amount of water. Salty water[f] is an electrolyte and, even if not the best, a conductor. Hence changing magnetic field applied to a meat chop will induce in it the Fucault currents as if it was of metal. The energy of electromagnetic field will transform to Joule heat and, as a result, the slice will roast.

A well known example of electromagnetic field changing both in time and in space is electromagnetic wave. But is hard to believe that whatever electromagnetic wave will be able to fry a steak. Bluntly irradiating it by flashlight one takes a risk to stay hungry. Therefore some criteria must be met. First of all, the field must be strong enough. For example, the field of a radar for watching aircraft would do. (Eyewitnesses tell that birds which by ill fate crossed the beam of a powerful radar fell dead not singed but rather boiled. Almost like in Baron Munchausen's tale.) Certainly for safety it would be nice to "confine" the field, localize and hold the wave.

Waves may be "stored" in resonators. For sound waves this may be a real wooden box. A body of violin is a typical resonator. Standing sound waves can last in it comparatively long after having been excited by external source. Naturally this role is played by the bow stick and strings.

Quite similarly it is possible to "store" electromagnetic field but the box must be metal. The length of the box must be equal to a whole number of halves of the confined wave[g]. Exciting (by means of some microradar) electromagnetic oscillations in the box turns it into a resonator with standing electromagnetic wave. The nodes of the wave (these are the points where the amplitude is zero) are at the walls. Microwave oven is exactly

[f]Remember the taste of bleeding finger...

[g]The same requirement is applied to sound resonators. —*A. A.*

such a resonator combined with a small microwave source. As the device is some tens centimeter ($\approx 1\,ft$) long we can readily estimate the maximal wavelength of the radiation inside. The check of the estimate may be found on the back of the oven. You will find that the standard frequency is $\nu = 2150\,MHz$, the corresponding wavelength being $\lambda = c/\nu \approx 14\,cm$.

Let us continue exploring the microwave oven and make an experiment as it once has been performed by the author himself. Take a bulky cut of deep frozen meat, salt and pepper it, put onto a special dish (its time will come yet), place into the microwave and switch that on. At first sight nothing happens but to the muffled hum of the ventilator (the necessity of that will be explained later). But then you will notice through the glass door that the meat will gradually become brown looking absolutely ready in thirty minutes. Take it out and cut in halves. It is quite probable that you will find inside a portion not simply raw, but even frozen. What is the explanation?

The simplest thing would be to blame a nonuniform field distribution in the oven. Indeed, the size of the compartment is about $30\,cm$ and the wavelength is ($\lambda \approx 14\,cm$). As the resonator is longer than four half-waves $\lambda/2 \approx 7\,cm$ the field inside must have at least three nodes where its intensity turns to zero. Spatial positions of the nodes of the standing wave are stationary and our piece of meat could have had a bad luck. Still in modern microwave ovens the problem is solved by slow rotating the table with the dish. This averages the effect of high-frequency field over the volume of the food. So, why not to take a merry-go-round for a turntable, load it with freshly extracted from eternal congelation mammoth and tug into a suitable microwave oven? What if the taste and the splendid amount of the outcome are worth effort?

Unfortunately an impassable obstacle thwarts the realization of the dream. It bears the name of ***skin effect***. This is the well-known property of high frequency currents to localize near the surface of conductor. As long as the frequency of the electromagnetic wave in the oven is very high this effect may be really important. The field will get damped in the bulk and the power will be not enough for frying.

In order to prove the last assumption let us dwell on skin effect. We shall try to estimate the effective depth of penetration of electromagnetic field and study its dependence on frequency and on properties of the conductor.

The questions can be easily replied by solving Maxwell[h] partial differential equations of electromagnetic field. Nevertheless this may scare some small minority of our readers and we shall recourse to qualitative estimates.

First, let us formulate the problem. Let an electromagnetic wave with frequency ω fall normally onto a flat surface of a conductor, Fig. 13.2. Inside the conductor the electric field of the wave drives electrons into motion. The currents lead to Joule losses and the wave dies out. It can be shown that the damping obeys the so common in nature (remember for example radioactive decay) exponential law:

$$E(x) = E(0)\, e^{-x\,/\,\delta(\omega)}. \tag{13.2}$$

Here $e = 2.71\ldots$ is the base of the natural logarithms; $E(0)$ is the amplitude of the electric field at the surface of the conductor and $E(x)$ is that at the depth x; $\delta(\omega)$ is the effective depth of the field penetration; that is the depth $\delta(\omega)$ where the field decreases e times.

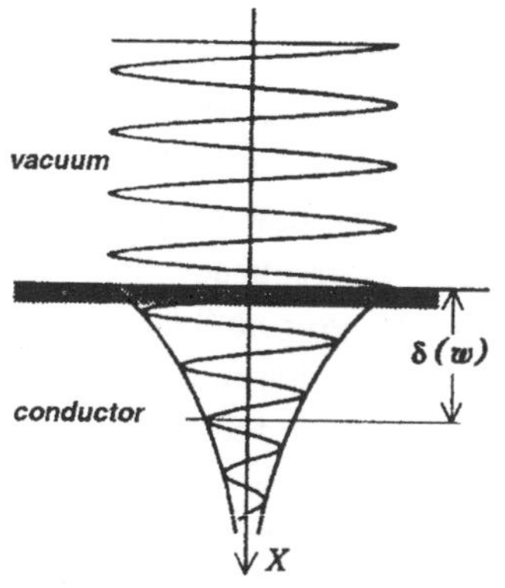

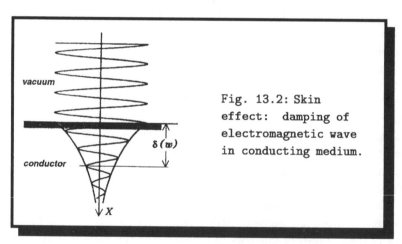

Fig. 13.2: Skin effect: damping of electromagnetic wave in conducting medium.

We shall find the value of $\delta(\omega)$ with the help of the dimensional method. Clearly the penetration depth must depend on the frequency. Remember that direct current ($\omega = 0$) flows through the full cross-section of conductor. So the skin effect is weak, $\delta \to \infty$, at low frequencies and becomes pronounced at higher ones. It is quite natural to suppose that the frequency

[h]J. C. Maxwell, (1831–1879), Scottish physicist, specialist in statistical physics, electro- and thermodynamics, optics *etc.*

dependence of penetration depth obeys a power law:

$$\delta(\omega) \propto \omega^{\alpha},$$

and expect α to be negative.

It is none less obvious that the penetration depth must depend on conducting properties of the sample. Those are characterized by the resistivity ρ of the material or, just the same, by the conductivity $\sigma = 1/\rho$. The essence of skin effect is that the energy of electromagnetic wave is converted to heat. The rate of Joule losses in a unit volume is:

$$p = j\,E = \frac{E^2}{\rho} = \sigma\,E^2,$$

where E is the strength of electric field and j is the current density at the point. (Try to derive this formula yourself. Remember that the local form of the Ohm[i] law is: $j = \sigma\,E$.) The more effective energy dissipation results into the faster damping. Thus the penetration depth must depend on the conductivity of the medium:

$$\delta \propto \sigma^{\beta},$$

and one may believe that β is negative like α is.

Finally let us note that the equations of electromagnetism when written in the International System of Units (SI) contain the dimensional magnetic constant $\mu_0 = 4\pi \cdot 10^{-7}\,H/m$, called the vacuum magnetic permeability. This constant enters into the expression for magnetic induction around electric current just like the vacuum dielectric permeability ε_0 appears in the formula for electric field of point charge. Let us assume that the penetration depth is a combination solely of these three parameters[j]:

$$\delta \propto \omega^{\alpha}\,\sigma^{\beta}\,\mu_0^{\gamma}, \tag{13.3}$$

and find the exponents α, β and γ by comparing the dimensions of the left and right sides of the equation. Write down the dimensions of all quantities

[i] G. S. Ohm, (1787-1854), German physicist; works on electricity, acoustics, crystal optics.

[j] This assumption implies that magnetic field is produced only by real currents. Neglecting the electric component of the wave (the so-called displacement currents) is well justified for frequencies in question. Otherwise the dimensional constant ε_0 would appear in the result.

in (13.3):

$$[\delta] = m, \quad [\omega] = sec^{-1}, \quad [\sigma] = \Omega^{-1} \cdot m^{-1}, \quad [\mu_0] = H \cdot m^{-1}.$$

Note that as *Henry* is the unit of inductance the relation

$$\mathcal{E} = -L\,\frac{\Delta I}{\Delta t}$$

makes possible to represent it as follows:

$$H = V \cdot sec/A = \Omega \cdot sec,$$

and

$$[\mu_0] = \Omega \cdot sec \cdot m^{-1}.$$

The two sides of the equation (13.3) must have the same dimension:

$$m = (sec)^{-\alpha}\,(\Omega \cdot m)^{-\beta}\,(\Omega \cdot sec \cdot m^{-1})^{\gamma},$$

or

$$m^1 = (\Omega)^{\gamma-\beta} \cdot sec^{\gamma-\alpha} \cdot m^{-\gamma-\beta}.$$

This is equivalent to the three equations:

$$\begin{cases} \gamma - \beta &= 0, \\ \gamma - \alpha &= 0, \\ -\gamma - \beta &= 1. \end{cases}$$

Solving those we find: $\alpha = \beta = \gamma = -1/2$. Substitution of these values converts the equation (13.3) into

$$\delta \propto \sqrt{\frac{1}{\mu_0\,\omega\,\sigma}}.$$

The dependence of the penetration depth on ω and σ confirms the preliminary physical analysis, since both α and β are negative. The accurate computation based on the Maxwell equations gives the same expression up to the numerical factor $\sqrt{2}$:

$$\delta = \sqrt{\frac{2}{\mu_0\,\omega\,\sigma}}. \tag{13.4}$$

Now we can estimate the penetration depth at the standard frequency $\nu = \omega\,/\,2\pi = 2.15 \cdot 10^9$ *Hz* used in microwave cooking. Conductivity of bulk meat is practically equal to that of muscle which, according to biophysical handbooks, is $\sigma_m \approx 2.5\,\Omega^{-1} \cdot m^{-1}$. It is interesting to compare it to the model conductor, copper, with $\sigma_c \approx 6 \cdot 10^7\,\Omega^{-1} \cdot m^{-1}$. The formula (13.4) gives for these values:

$$\delta_m \approx 1\,cm \qquad \text{and} \qquad \delta_c \approx 10^{-3}\,cm.$$

The effect is rather strong for good conductors and leads to enhancement of resistance of wires at high frequencies. The analysis proves that alternating currents are concentrated in the layer of the thickness $\delta(\omega)$ near the surface. At high frequencies δ is small and the effective cross-section of wire falls down. However even for such a bad conductor as meat the effect is quite noticeable. Our test (10 *cm* thick) beef cut turned out to be too large. The field strength and the released heat diminish many times in the center and thermal conductivity becomes the only source of heat. And turning back to the mammoth, here the skin effect shows up mockingly literally. At best only the thick hide of the giant would fry whereas the meat would remain untouched.

Well, one can survive that. Just don't put into your oven too thick slices and meat will roast safely and thoroughly. A big but flat steak is not sensitive to skin effect. Maybe you will even invent some delightful applications of the skin effect. For example, try to fantasize sometime about preparing in a microwave such an exotic dessert as... ice cream in hot freshly baked pastry crust[k].

Like everything in the world the microwave oven has its drawbacks. On the one hand it opens prospectives of pioneer cooking dishes that nobody has ever heard of, it preserves vitamins and provides a means of making dietary products but on the other it is incapable to imitate a soft-boiled egg. Indeed, let us **imagine (!)** an egg placed in the oven. After the power has been switched on the Fucault currents emerge in the liquid contents.

[k]There is a mechanism that facilitates the task. As you know, electric current in water is a motion of ions. This means that the conductivity of ice σ_- where ions are fixed by crystal lattice is less than that σ_+ of water. This increases the penetration depth, $\delta_- \propto \delta_+ \sqrt{\sigma_+ / \sigma_-}$ but reduces the Joule heat: $p_- = \sigma_- E^2 = p_+ \sigma_- / \sigma_+$. Hence the heat release in the frozen region must be suppressed with respect to the covering pastry. So if the ice cream is fluffy (air bubbles add to dielectricity) and well cooled the chances that it will bear up the baking are not so bad.

The same mechanism worked in the frozen meat lump where it kept the inside icy and raw. The reduced power program "defrost" helps to preserve surface layers from getting ready too fast. —*A. A.*

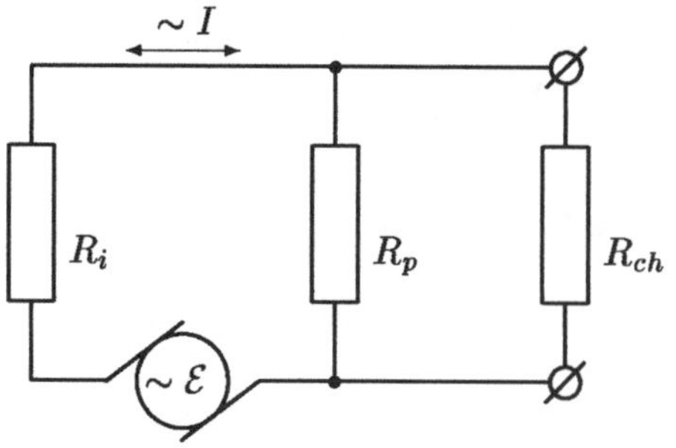

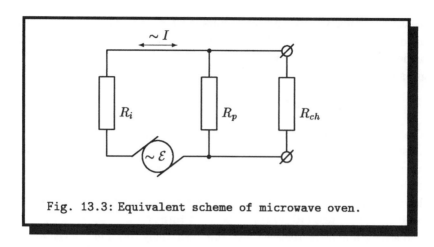

Fig. 13.3: Equivalent scheme of microwave oven.

The fast heating causes release of vapor which can not get out of the shell. The pressure grows precariously, one more second and... BLAST! The egg blows up. The whole compartment has gotten splashed with the anticipated breakfast and you passionately disparage the lack of foresight while scrubbing it.

Now, after the relatively harmless explosion, it's right the time to talk about safety. Remember that the maximum power of microwave ovens is rather high being about $1\,kW$. The power can be regulated but still there must always be something inside where the heat could be dissipated. Therefore it is strictly forbidden to operate an empty oven. Missing the object to act upon the high frequency field "starts searching" where it could dump the energy and turns the firepower against itself. The heat will be released in the emitting elements and destroy them.

In order to illustrate what happens let us look for analogy. Suppose that we placed into our microwave a pretty seasoned chicken. The physics of frying process is reflected by the following equivalent scheme, Fig. 13.3. A source of alternating voltage with the electromotive force $\mathcal{E}$ (that imitates the microwave generator) with the internal resistance R_i is closed by a parasite resistor R_p (in our case this role belongs to emitting elements and walls of the oven) and connected to some external load R_{ch} (probably you have already guessed that *ch* stands for the chicken). No doubt engineers do their best in order to maximize R_p and minimize the losses. But although $R_p \gg R_i$ it is impossible to make it infinite. Inasmuch as R_p and R_{ch} are connected in parallel the net resistance is $R_n = R_p \cdot R_{ch}/\,(R_p + R_{ch})$.

The current in the circuit is $I = \mathcal{E}\,/(R_n + R_i)$ and the power spent by the source $P = \mathcal{E}\,I$ is divided between the three resistances and converted to heat. The optimal working regime corresponds to $R_{ch} \approx R_i \ll R_p$ and the heat released in the circuit is distributed as follows:

$$P_i^{\text{opt}} \approx P_{ch}^{\text{opt}} \approx \frac{P_0}{2} = \frac{\mathcal{E}^2}{4R_i}; \qquad P_p^{\text{opt}} \approx \frac{\mathcal{E}^2}{4R_p} = P_0 \frac{R_i}{2\,R_p} \ll P_0;$$

here P_0 is the power consumed under the optimum conditions. Note that one half of the energy (up to $0.5\,kW$) is dissipated in the source itself. This explains why ventilator is the indispensable detail of microwave oven.

Now, what happens if bad boys have replaced the true chicken by a plastic copy with $R_{ch} \to \infty$? (This is almost the same as leaving the oven empty.) The total energy consumption will substantially fall down and the power will be shared as follows (∞ stands for the dielectric chicken, $R_{ch} = \infty$):

$$P_i^{\infty} \approx \frac{\mathcal{E}^2\,R_i}{R_p^2} = 2P_0 \frac{R_i}{R_p} \ll P_i^{\text{opt}} \qquad \text{and} \qquad P_p^{\infty} \approx \frac{\mathcal{E}^2}{R_p} \approx 4\,P_p^{\text{opt}}.$$

Note that the parasite heat production has increased four times and the greater part of it falls on the emitters. Hardly getting them fried is desirable.

Operating empty microwave oven is fatal for the emitting elements.

But a much more serious danger is concealed in choosing wrong utensils. Of course some prefer to bye special glass-ware *"Pyrex"* and use it. But we recommend to think over physical requirements to kitchenware, take an old earthen pot and save the money. The main demand is that the pot must be transparent for the microwave radiation. It must remain dielectric even at high frequencies. Electric and magnetic properties of the material can markedly depend on the frequency of electromagnetic field. So by far not every glass or faience serves the purpose. And not under any pretext neither metal pots and foil-wrapped foods nor even gold-rimmed china must be put into the oven. In the twinkling of an eye the sympathetic kitchen accessory will turn into its fire-spouting relative from the founding shop and wreak havoc in and outside.

Let us turn to the equivalent scheme in Fig. 13.3 once again. Suppose that grandma stuffed a chicken with prunes and chestnuts, sprinkled it with ground cloves an cinnamon, then arranged all on the treasured cast iron pan and put that into a microwave. Because of the pan the effective resistance of such a masterpiece is zero, $R_{ch} = 0$, and it will short the current source. The current will bypass the parasite resistance and take the short-circuit value, $I_0 = \mathcal{E} \,/\, R_i$. The frying pan will get no heat, $P^0_{ch} = I_0^2 \, R_{ch} = 0$ and power will dissipate only in the microwave source:

$$P_i^0 = \frac{\mathcal{E}^2}{R_i} = 4P_i^{\rm opt}.$$

You see that now the heat released in the electric part exceeds the standard value four times and reaches $2\,P_0 \approx 2\,kW$! No chance, good if the device alone will be fused.

Metal in the chamber of microwave oven short-circuits the high-frequency generator.

Nevertheless a lot of earthen pots and ceramic plates will do. To make sure whether a utensil fits microwave cooking you may put it into the chamber along with a glass of water (what for?) and turn the switch on. If two minutes later the object of testing remained cool leave your worries, it had passed the exam.

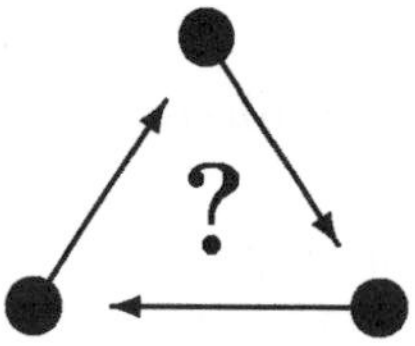

Try to estimate the depth of penetration into brain of the radio waves emitted by a cellular phone ($\nu =$ 900 MHz).

Chapter 14
The water mike

About one invention of Alexander Bell

These days, everybody knows what the microphone is. Right? We still quite frequently see it on TV: those nifty pin-looking things intricately woven in the anchormen's lapels, or the rather old-fashioned, a ball or a rod-with-handle type, reporters stick into people's faces in feverish anticipation of a day-making news; radio-interviewers often ask their guests on the program to speak directly into their microphone hinting to us the presence of the latter; the movie makers, no matter how sophisticated and outlandish their sound effects are contrived to be, end up registering them with some kind of a microphone. One can easily buy a decent mike in any "Radio Shack" or "Best Buy" store and use it with her tape or CD recorder, computer or telephone. The design of the gadget is described in most of the today's high school physical textbooks. However, we can assure our revered reader that there are only a few who are aware of the existence of the so called water microphone. Don't get surprised. Indeed it turns out that one can rather efficiently amplify different sounds with the help of a simple water streamlet. The device employing such a principle of sound amplification was invented by the American engineer Alexander Bell, who is mostly known as one of the inventors of another gadget we can't imagine our daily life without — the telephone[a].

But at first, let's pay attention to that "water stream" amplifier thing.

[a]Alexander G. Bell, (1847–1922), Scottish-born American inventor. The first public demonstration of the speech transmission using his electrical apparatus took place in 1876.

If there is a hole, say a little round orifice, drilled in the bottom of a reservoir with water, one can notice that the stream flowing downwards through the hole consists actually of two, differing in their properties, parts. The upper one is transparent and steady looking as if made of glass; yet as it goes further from the outtake, the stream becomes thinner and thinner and finally at the point of the minimal cross-section, it turns into the second part, which is rather opaque and jittery. At the first glance, it still looks continuous, without interruptions, like the upper region. However, it turns possible at times to swiftly pass one's finger across this part of the stream without even wetting it. The French physicist Felix Savart[b] after having meticulously investigated properties of liquid streams, arrived at the conclusion that at the narrowest point of the stream, it breaks its continuity and splits into series of separate droplets. Today, over century later after the discovery, one can easily prove this by taking photographs of the trickle with flash, or by looking at the stream in stroboscopic lighting Fig. 14.1; in those old times, however, researchers had to study the trickles in the dark, observing them with light from electrical spark.

Look at the momentarily image of the lower portion of the stream, Fig. 14.1. It's composed of successive, alternatively bigger and smaller, droplets. As the picture clearly shows, the bigger ones are actually oscillating, gradually changing their shape from a flattened, stretched horizontally ellipsoid (droplets 1 and 2 in the picture), to round balls (3), then to the ellipsoid again (4, 5, 6), yet now squeezed and stretched vertically, and then back to sphere (7), *etc., etc.* Each droplet, pulsing rapidly in its free fall[c], produces different images in one's eye at different instants. This causes that perception of the lower part of the stream as kind of misty, widening in the regions where the droplet-ellipsoids are stretched latitudinally, and inversely, — narrowing where they are elongated vertically.

[b]F. Savart, (1791–1841), French physicist; works in acoustics, electromagnetism and optics.

[c]The frequency of pulsations can be estimated similarly to that of air bubbles in liquid, chapter 9, "The chiming and silent goblets":

$$\nu \sim \sigma^{1/2} \rho^{-1/2} r^{-3/2}.$$

Putting $\sigma = 0.07\, N/m$, $\rho = 10^3\, kg/m^3$, $r = 3 \cdot 10^{-3}\, m$, one finds that $\nu \approx 50\, Hz$. Worth noting here is that the "shooting rate" of a typical movie camera, is 24 pictures per second. It happens to be already enough for the human eye to take the film for continuous motion.

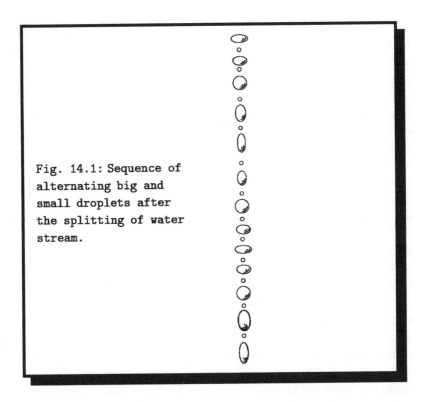

Fig. 14.1: Sequence of alternating big and small droplets after the splitting of water stream.

Another interesting finding of the French scientist was that there is a strong effect of the surrounding sounds on the upper transparent part of the water jet: if a sound of a certain pitch (meaning frequency) is excited nearby, the transparent region of the stream turns instantly opaque. Savart gave it the following explanation. The droplets, in which the stream finally breaks down in the bottom part, start developing actually from the very beginning of the fall, right by the outlet orifice. At first they are outlined just by sort of circular notches that become more and more prominent as the liquid falls, until the point of the stream where they split completely. These notches are so close following each other, that they make a slight sound. Thus a music tone, in unison with this "natural" pitch, will make the continuous stream of liquid break into separate droplets earlier turning the transparent flow bleary.

The English physicist John Tyndall[d], repeated later the F. Savart's

[d]J. Tyndall, (1820–1893), English physicist, specialized in optics, acoustics and mag-

experiments in his laboratory. He had managed to produce a water stream, transparent and uninterrupted, of about 90 feet long. And then, by using the sound of appropriate tone and volume from one of the pipes of an organ, he was able to break this stream into countless separate drops, transforming it by doing so into a misty unsteady trickle. In one of his articles, he related his observation of the water stream falling into a basin. He observed that of the water stream falling into a basin sounds like that when the falling stream crossed the surface above the point of transition from the transparent to opaque part at moderate pressure, then the stream entered the liquid silently; but when it crossed the surface below the rupture point, a murmuring started and one saw numerous bubbles formed. In the former case, not only a serious sputtering did not take place, but the liquid rater piled up around the basis of the stream in the basin where the direction of motion of the liquid was actually reversed.

The described above features of water streams were used by A. Bell in his design of the water mike, depicted in Fig. 14.2. It consisted of a metal tube and a branch pipe with a funnel soldered on its side; the bottom of the tube was mounted on a massive support, whereas its top was covered with a membrane-like rubber piece, fastened to the tube by a lace. As we know already from the Tyndall's experiments, the lower part of the water stream, split into separate droplets, makes a murmuring noise as it reaches the water in the basin. On the other hand, when the upper, still continuous, portion of the stream enters the liquid in the reservoir, the flow remains soundless. One can perform a similar demonstration with a piece of cardboard, placed across a stream of water. As one pulls gradually the cardboard sheet up, the drumming generated by hitting droplets becomes lighter and lighter and, after the "transition point" is passed, the noise ceases at all.

The membrane in Bell's microphone plays exactly the same role as the cardboard piece in the preceding example. Yet now, because of the resonator (tube) and the side pipe with the trumpet, any slightest tap of droplet is amplified and echoed much louder. Thus, the tiny droplets when hitting the rubber membrane will make the rapping worth that of a hammer against an anvil.

One could easily use this apparatus to illustrate the sensitivity of water stream to different musical tones, the fact described by Savart and Tyn-

netism.

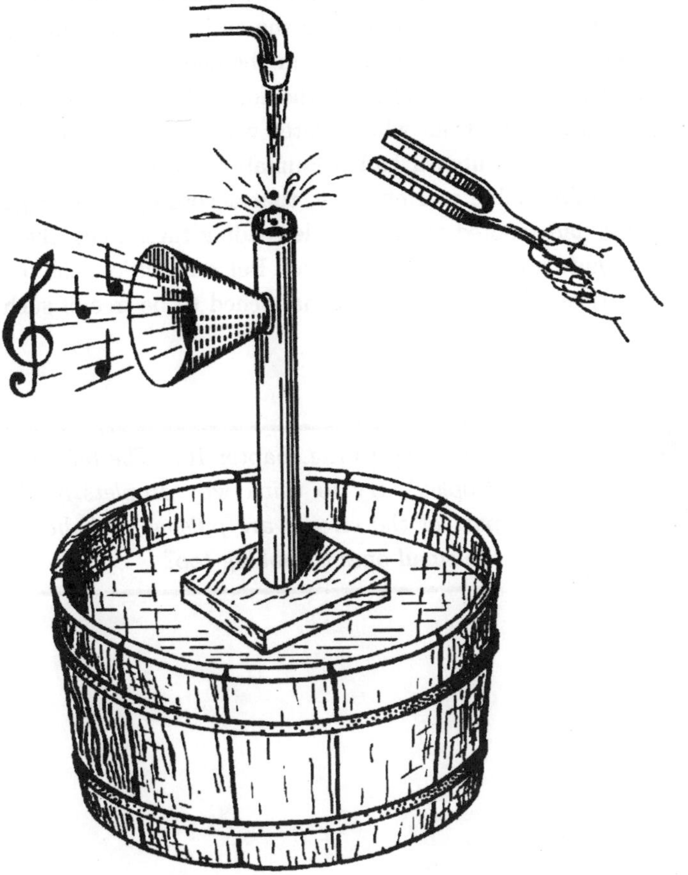

Fig. 14.2: The water microphone by Alexander Bell amplified sound at the expense of energy of falling droplets.

dall. That is, if we touch by vibrating tuning fork the outlet of a faucet from which a slim stream of water runs, Fig. 14.2, the flow will momentarily break into drops which "commence" their rather earsplitting chorus. This amplification of the originally pretty weak sound at the expense of the energy of falling stream does indeed constitute the physical principle of the water microphone. If one substitutes the tuning fork with, say, a wrist watch, it will make its ticking audible to the entire audience in the room. One of renowned popularizers of science in the end of nineteenth century, claimed that he had tried to transmit the sound his voice by connecting a funnel to a glass tube from which water was running. And, although the water stream in his utensil had presumably started "talking", but in so horrifying indiscernible roaring voice, that according to the legend the spectators just had scooted away[e]. While reading these lines, the authors feel that it is quite a blessing that the main Bell's invention, the telephone with the electrical mike in its receiver, happened to be free of such a disadvantage.

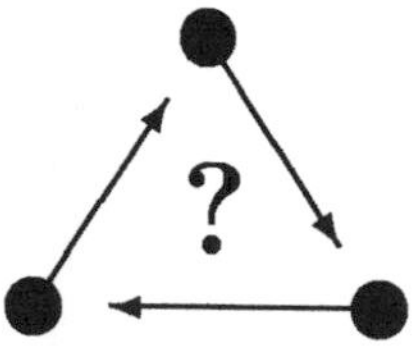

Returning to the chapter 10, "The bubble and the droplet", try to think why droplets in the lower part of the stream are pulsing periodically? Are the small ones throbbing too?

[e]See footnote on page 70.

Chapter 15

How the waves transmit information

Recently, we have grown so accustomed to the television, radio, cellular phones and Internet, that we are not any more a tiny bit surprised by the fact that we can so easily receive the needed information from pretty much any corner of the world. Yet it was not always like that (actually not even close) and not a very long time ago:

It's a Russian book after all so we may just as well use an example from Russian history. In order to send a message to St. Petersburg about the coronation ceremony of the queen Elizabeth which took place in Moscow in 1741, all way along the road connecting these two cities, were mustered and aligned in a kind of human chain a couple of thousands of soldiers with signal flags in their hands. When the crown was put on the head of the new empress, the first of the soldiers waved his flag, so did the second when he had seen his neighbor signaling, then the third, the forth and so on. So the news about crowning event had made its way to the Northern Russian Capital, where a fired cannon notified the crowds of anticipating subjects.

Now, let's ask ourselves a legitimate question: what was actually "moving" along this peculiar chain? Although each soldier remained in the original place, yet at certain moment he changed his ***state,*** raising the flag. And this change of state was exactly the thing which moved along the chain. In situations of this sort physicists say that a ***wave was running (or propagating)*** along it.

There is a great variety of different kinds of waves, depending on what actual physical property does vary when they propagate. For the sound (acoustical) waves, the density of matter they run through fluctuates, whereas, for instance, in the electromagnetic waves (light, radio, television, etc.), the intensity of electric and magnetic fields are oscillating. There are temperature waves, waves of concentration in chemical reactions, epidemic waves and so forth and so on. In a poetic word, one could say that waves penetrate the entire edifice of the contemporary science.

The simplest possible type of waves are the monochromatic ones, when states change at each point according to the simple harmonic law, with certain constant frequency, (the sine or cosine law). The monochromatic sound waves are what we call the sound tones, or pitches. One can excite such waves using, for example, the tuning fork. The monochromatic light waves are generated by "their majesty" lasers. With a simple stick, while dipping it periodically up and down in water, one can make pretty close to being monochromatic ripples. The similar waves could be produced in our live chain as well.

Imagine that each grenadier not just simply raises his flag, but rather starts continuously and periodically waving it from side to side; and each next soldier follows the one before him, yet with a certain fixed delay or *phase shift*. A wave takes off running along the chain. We bet, our dear reader has seen something like that at a stadium during competitions, when the ardent (or just simply bored) spectators start making what is called "live wave".

These monochromatic events are quite pleasing to the eye, yet can they actually transmit information? — Obviously, no. Periodic in time oscillations do not tell us anything new, so no information is transmitted. Whereas, with the single toss of hand the diligent servicemen did manage to communicate to St. Petersburg (more than 600 *km* away from Moscow) the important news. What is the difference between these two kinds of waves? If we take an instant pictures of the motion for both situations, we will see that in the first case all soldiers are involved in the motion, whereas in the second — only one of them. In other words, when a signal is being transmitted, the wave (whatever kind it was) at each moment is localized in space. One could imagine two, or three, or even several near standing participants raise their hands at the same time. In such a case, the length of conveyed signal would increase. Having been able to generated signals of different length, one could send not only message about a single event (such as the coronation having been accomplished, for example) but, in principle, any kind of information. Take, for instance, the famous Morse code (which was patented much later by the way, in 1854)[a].

Besides of regiments of grenadiers, there are, of course, other information-transmitting signals: light, sound, electric current, *etc.* It's interesting to note that any signal can be presented as a sum of the monochromatic waves with different frequencies. This possibility is ensured by the so-called ***superposition principle,*** stating that the oscillations of overlapping (interfering) waves at each point of a medium, should be just simply summed up. And, therefore, depending on the phase shift, oscillations can either amplify each other (for example, two identical waves with no phase shift will result in the oscillation of doubled amplitude, Fig. 15.1, *a*) or extenuate each other (again: two identical waves being out of phase will cancel each other completely, Fig. 15.1, *b*). It turns out also, that one can opt (or tune) the amplitudes and frequencies of combined monochromatic waves in such a manner that they will nullify each other in the entire space, except for just a certain region where, conversely, they will amplify each other.

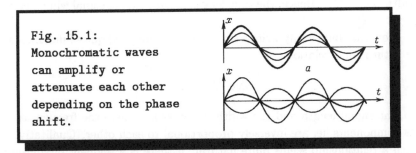

Fig. 15.1: Monochromatic waves can amplify or attenuate each other depending on the phase shift.

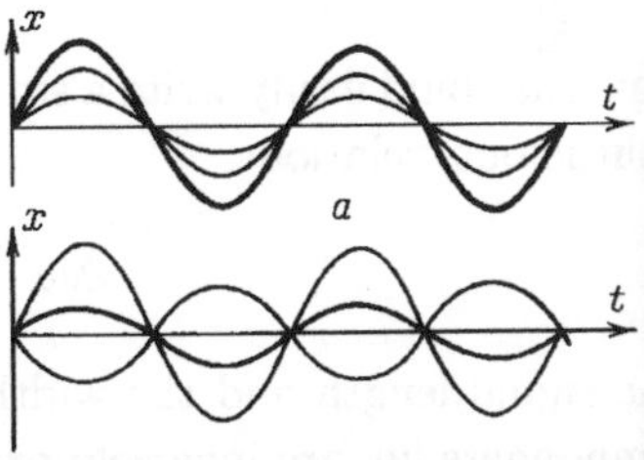

The result of summation of a large number N of waves with identical amplitude A_0, and frequencies lying within a small interval $2\,\Delta\omega$ around the basic frequency ω_0 is shown in Fig. 15.2. It is like an instant photograph of wave, showing variation of a fluctuating quantity A in different points of space at a fixed moment of time. There is a central maximum with $N\,A_0$ amplitude, and also numerous secondary peaks, though with quickly waning amplitudes, indicating that, indeed, the overlapping waves mostly extenuate themselves, having been noticeably amplified only in the vicinity of the central maximum.

[a]Surprisingly, but the marine semaphore system where various positions of the two, red and yellow, flags meant different letters and figures appeared much later, in 1880. —*A. A.*

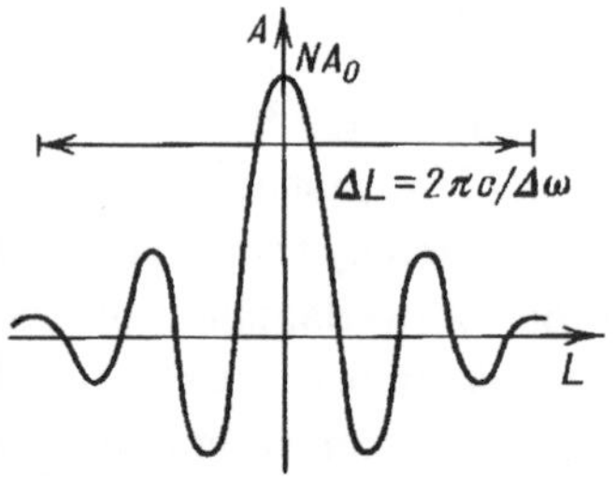

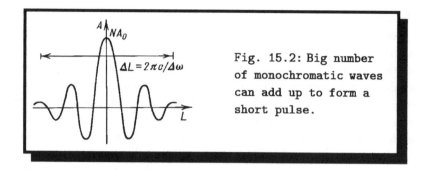

Fig. 15.2: Big number of monochromatic waves can add up to form a short pulse.

Another worth noticing thing about the superposition is that the central peak does not stand still, yet is moving with the wave's propagation speed. If the monochromatic component waves all move with the same velocity c (like in the case of electromagnetic waves in vacuum, for instance), the central maximum also moves with the same velocity c, keeping its constant width $\Delta L = \frac{2\pi c}{\Delta\omega}$. Hence, the time length of the running signal will be equal to $\Delta t = \frac{2\pi}{\Delta\omega}$.

Now one could easily write a simple yet amazing and quite fundamental, as it turns out, relation:

$$\Delta\omega \cdot \Delta t \sim 2\pi.$$

So, the signal length and the width of the range where the frequencies of its components lie, are inversely proportional to each other. Qualitatively, such a relationship seems quite natural: if there is a segment of sinusoid, corresponding to a signal of rather long duration (Δt is large), then it must be an almost monochromatic wave and $\Delta\omega$ is small. However, if a short signal is required, it should be combined of many waves with different frequencies. Everyone, we believe, noticed the glitches and noises generated in the radio for pretty much all bands, when a lightning strikes nearby.

Thus, each signal can be made of a set of monochromatic waves, or, just the same, each signal can be decomposed into such waves. The amplitude versus frequency dependence for monochromatic waves, composing a signal, is called the signal ***spectrum***[b]. In the presented situation, for example, it will be a rectangle of altitude A_0 and width $2\,\Delta\omega$, depicted in Fig. 15.3. This,

[b]Although physicists sometimes mean by spectrum just the set of frequencies of monochromatic waves, making a signal, we stick to the more particular definition that takes into account the waves' amplitudes as well.

of course, is a trivial spectrum; signal spectra, just like signals themselves, can have various, most peculiar at times shapes.

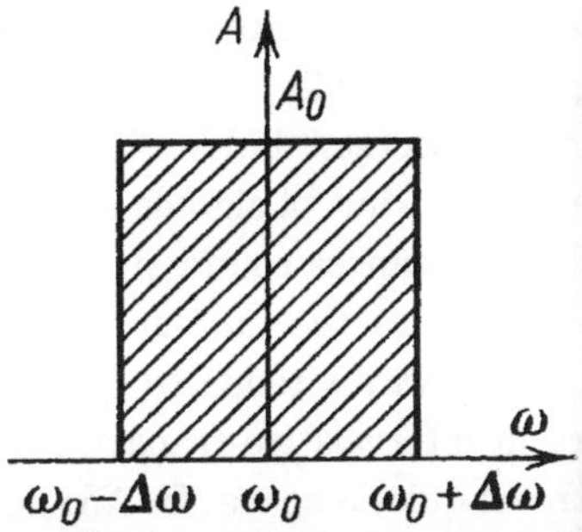

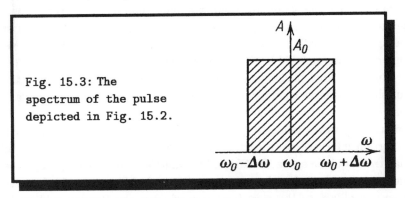

Fig. 15.3: The spectrum of the pulse depicted in Fig. 15.2.

When we pronounce sounds, for example, we make the air vibrate in a certain way, so these vibrations propagate as sound signals of certain shape. Their spectra strongly vary depending on whether we utter vowels or consonants. The vowels have spectra with two distinct characteristic peaks at certain frequencies (they are called ***formants***). The spectra of consonants, on the other hand, are more "smeared", spread over the whole audible sound frequency range: Fig. 15.4 represents the spectrum of the letter $\mathcal{S}$ sound. There has been developed an entire method called ***harmonic analysis*** allowing to find spectra of registered signals, as well as restore signals from their spectra.

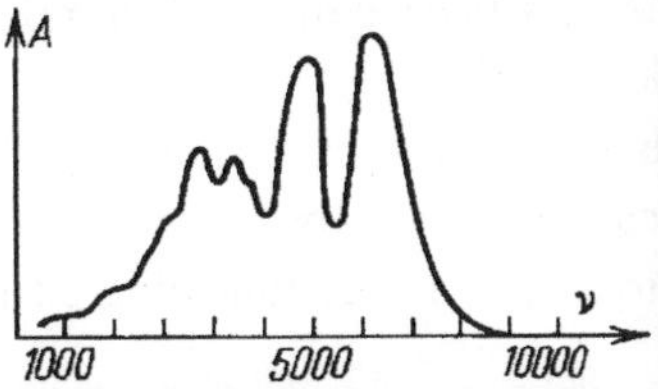

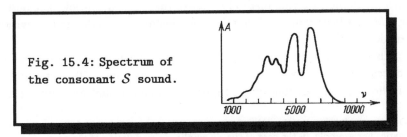

Fig. 15.4: Spectrum of the consonant $\mathcal{S}$ sound.

Peculiar though it may sound, the solid bodies are capable of "yelling" too. The thermal motion makes atoms in the crystal lattice oscillate, generating, thereby, elastic waves propagating inside the crystal. These oscillations, are sound waves too. Their spectral maximum, however, lies in the area of extremely high frequencies — at the absolute temperature of about $5^\circ K$ it stays around $10^{12} - 10^{13}\,Hz$. In the audible frequency range, though, the amplitude of these oscillations is negligible; so that in order to hear "what the solids are talking about", one must use some special devices. And by "listening" to this "chatting" (studying the sound signal spectra, actually), researchers have already discovered plenty of very important "secrets" hidden in the solid-state.

But what kind of signals are used in practice to transmit information? For short distances sound signals work just fine, and for a long while, already, as far as the human history goes. The limitation though is that this type of waves tends to quickly dissipate. Yet, if one amplifies them at certain intervals (re-transmits them), sending them along, such signals may travel quite a space. In Africa, for example, until recently, people were sending messages using the tam-tams, "drumming" a piece of information from one village to the next (just sort of what the Russian soldiers had done with their flags).

In the contemporary world, however, most of the signals are transmitted in the form of electromagnetic waves, which can cover much longer distances, before dying out. One, for example, can make an electromagnetic wave carry sound signals. In order to do that, the frequency of this electromagnetic wave (called the *carrier wave*) is kept constant, whereas its amplitude is varied (*modulated*) in accordance with the sound oscillation to be transmitted, Fig. 15.5. This way a signal containing the needed information is generated. Then, at the receiving end, the signal is "deciphered",— the envelope corresponding to the modulating sound signal is extracted. Such a sending-receiving method is called, therefore, ***amplitude modulation*** or *AM*. It's been employed in galore, for instance, in radio and television broadcasting[c].

There arises a next question though: how much information per unit time can one actually transmit, with the help of waves? To sort this one out, let's look at the following situation. It's known that any number can

[c]Of course, after modulation, the electromagnetic wave remains no longer monochromatic. For example, in the case of simple amplitude modulation of the carrier wave of, say, frequency ω_0 and amplitude $A(t) = A_0\,(1 + \alpha \sin \Omega\, t)$, in Fig. 15.5:

$$x(t) = A(t) \sin \omega_0\, t = A_0 \sin \omega_0\, t + \frac{\alpha A_0}{2}[\cos(\omega_0 - \Omega)\, t - \cos(\omega_0 + \Omega)t]$$

You see that the spectrum even of this simplest modulation consists already of three different frequencies: $\omega_0 - \Omega$, ω_0 and $\omega_0 + \Omega$.

Fig. 15.5: Amplitude of amplitude-modulated carrier wave varies in accordance with the transmitted low-frequency signal.

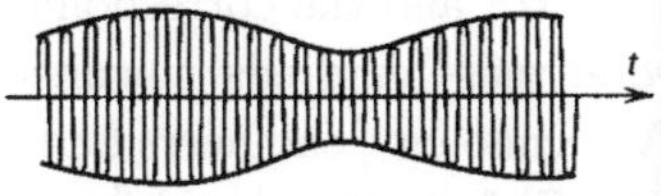

be presented in the binary notation, as a sequence of ones and zeros. In the similar way, any information can be written encoded into a row of successive pulses and pauses of certain duration. The signals can be transmitted by amplitude modulated waves, for instance, Fig. 15.6. The higher the speed of information transmission is wanted, the shorter these signals must be. Yet for reliably transporting the information the length of the signal shouldn't be shorter than the period of the carrier sinusoidal wave. This gives us right away the categorical limit on the maximum rate of information transmission. When one wants to raise the speed, one necessarily needs to increase the carrying frequency. Here, indeed, once again tells the considered relation for the time length of a signal: $\Delta t \approx \frac{2\pi}{\Delta\omega}$, where $\Delta\omega$ becomes comparable to ω_0.

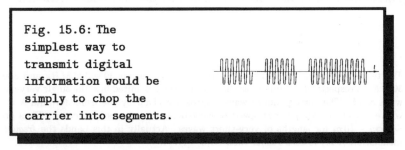

Fig. 15.6: The simplest way to transmit digital information would be simply to chop the carrier into segments.

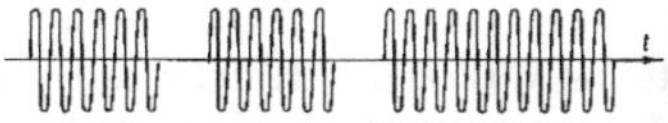

For example, to broadcast a musical program, it is sufficient to use electromagnetic waves with frequencies of around several hundred kilohertz ($1\,kHz = 10^3\,Hz$). The human ear is capable of sensing sound frequencies up to approximately $20\,kHz$, and, hence, the frequencies composing the signals, in this case, will be at least an order of magnitude less than that of the carrier wave. Yet, for transmitting television programs, this frequency range doesn't work already. The images on the TV screen are generated 25 times per second, and they, in turn, consist of tens of thousands of separate dots (pixels). So the needed modulation frequency is about $10^7\,Hz$ and the corresponding carrying wave should have frequency of several tens–hundreds megahertz ($1\,MHz = 10^6\,Hz$). That's why in the TV technology, engineers had to utilize the ***very-high (VHF)*** and ***ultra-high (UHF)*** frequency bands[d]. Consequently, the ***very-short*** radio waves, with wave length of the order of one meter, are employed even though such waves can propagate only over rather limited (basically, within the direct visibility) distances[e].

An even better candidate for the fast information transmission would be the ordinary light, having its frequency in the area of $10^{15}\,Hz$, which would allow to boost the transmission speed by several tens of times. Although the idea is rather old (Alexander Bell[f] actually was the first who applied light signals for sending sound messages back in 1880), its technological feasibility has arrived only recently. The success added up of the development of the high quality sources of monochromatic light — lasers, coupled with the fiber-optical light guides, capable of transmitting light with extremely low propagation losses, as well as all grandeur of the modern electronics, allowing efficient coding-decoding of these light signals.

Now we can with all certainty state that the age of "copper wires" is fading away and is being substituted by the coming epoch of ultra fast information transmission networks based on the fiber-optics technology.

[d]The terms VHF and UHF refer to the frequency (ν) bands $30 - 300\,MHz$ and $300 - 3000\,Mhz$ respectively. Sometimes the both ranges are united into the single very-short wave band. The corresponding wave lengths are given by the usual formula $\lambda = \frac{c}{\nu}$, where $c = 3 \cdot 10^8\,m/s$ is the speed of electromagnetic waves in vacuum. This gives $\sim 1\,cm - 10\,m$ for lengths of very-short waves. Actually in this bands the ***frequency modulation (FM)*** is used rather than AM. —*A. A.*

[e]Interesting enough, that the first television sets (with mechanical field scanning), made back in the twenties of XX century, worked in the medium wave (MW) band. The quality, due to the discussed problems, was very poor and the images were almost indiscernible. That had made researchers and engineers switch to very-short waves and develop the electronic field scanning techniques. However, this MW television had its own advantage — because of the longer (compared to the very-short waves) propagating range, programs from Moscow, for example, could reach, say, Berlin without any TV satellites or retranslating stations.

[f]Whom you already know from Chapter 14.

Chapter 16

Why the electric power lines are droning

> What is the Tacoma Narrows Bridge? The Tacoma Narrows Bridge was built in 1940. After Months of swaying up, down and side to side the Bridge collapsed. Taking with it the life of a poor dog (Tubby).
>
> Mr. H's world O'Physics.

Long times ago, the Ancient Greeks had noticed that tightly drawn string began sometimes to sound melodically in the wind, like if it was singing. Perhaps, back then people already knew the Aeolian harp, named so after Aeolus, the God of Winds in the Greek mythology. It is made up of a frame (or an open box) with several strings stretched across; it is put, then, in a place where the wind could pass through it. Even a single string in such an instrument can generate quite a spectrum of different tones. Something of the same nature occurs, although with far more limited tune variety, when the wind swings the cables of telegraph lines.

The puzzle of this phenomenon had for long bewildered scientific pundits of the past, until, in the end of seventeenth century, Sir Isaac Newton had applied his, back then newly developed, analytical method to problems of what we would call today fluid dynamics.

According to the law, first stated by the great Englishman, the resistance force acting on a body moving in liquid or gas is proportional to the square of the velocity v:

$$F = K\rho v^2 S.$$

Here S is the area of the perpendicular to the direction of motion cross-section of the body, ρ is the density of liquid (or gas), and K is just a proportionality coefficient.

Later though, it turned out that the formula does not apply universally. When the speed of the body is low compared to thermal velocities of molecules of the medium, the above relation starts faltering. We have discussed already in Chapter 11 that for relatively slow moving bodies the resistance force becomes linearly proportional to the speed (Stokes' law). This situation occurs, for example, when tiny droplets move in a rain cloud, or residue flakes precipitate onto the bottom of a glass, or the drops of substance $\mathcal{A}$ are roaming restlessly inside the *magic lamp* (see the same Chapter 11). However, in the modern world with its jet velocities the Newton's law for resistance force is much more common.

Either way, couldn't we find a satisfactory explanation to the phenomena of droning power lines or the tunes of the Aeolian harp just knowing these basic relations for the resistance? The answer is — no. Unfortunately, it is not at all that simple. Really, if the resistance force would remain the same (or was increasing with the speed rising), the string would be drawn by the breeze more and more, producing no oscillations at all.

So, where is the trick here? Well, it turns out that in order to understand the nature of the string vibration in this case, we can not get away with just a couple of rather general, not touching upon the flow mechanism, ideas. We ought to delve a bit deeper and discuss in more detail how does liquid actually flow around a body resting in it. (This is simpler than considering a body moving in the static liquid but does not affect the result, of course).

The case when the current is relatively slow is depicted in Fig. 16.1. The liquid flow lines are passing smoothly around and behind a cylinder (the picture shows the cross section). The flow of this kind is called *laminar*, and the resistance force arises from the internal friction (viscosity) of the liquid and indeed is proportional to velocity of the liquid (again — our reference frame is tied to the body). Both velocity and friction force at any point in such a flow are time-independent (the flow is ***stationary***), and this rather insipid situation is of no interest to our Aeolian problem.

Now let's look at Fig. 16.2. The flow velocity has increased and new whirling characters have appeared — the eddies or vortices, if you wish. Friction is not the definitive cause any more. Now it depends less and less on changes of momentum on microscopic scale, yet scales comparable to the size of the body itself start dominating. The resistance force becomes

Fig. 16.1: Lines of slow laminar flow around long cylindrical wire.

proportional to the second power of velocity, v^2 .

Fig. 16.2: At higher flow velocities vortices appear behind the wire.

Finally, see Fig. 16.3, the flow velocity has grown up even more, and the eddies are now aligned into neat well-ordered chains. And here indeed lies the answer to the riddle of string's vibrations! The thing is that these nicely structured tails of vortices break loose recurrently from the surface of the string and thereby do excite the string oscillations, like plucking fingers would do.

Fig. 16.3: In fast flows a periodic vortex trail is formed in the wake.

This, quite peculiar at first glance, arrangement of vortices tailing behind the body, had been first discovered and studied experimentally in the beginning of twentieth century. It found theoretical explanation in works of the Hungarian scientist, Theodore von Karman[a]. Now these periodic vortex wakes are known as the ***Karman trails*** (or even the Karman Vortex Street).

As velocity continues to increase even further, the vortices do not have enough time any more to spread over a large area of the liquid. The "swirling" region narrows, the eddies start mixing with each other and the flow becomes chaotic, irregular (***turbulent flow***). The latest experiments, though, show that for extremely high velocities there develops another kind of periodicity, but here we are getting well beyond the scope of our essay, and the best we can do for the curious ones is to refer them to, say, the exiting book by James Gleick, titled "Chaos".

It's worth noting that, although the discussed here phenomenon of Karman vortex trails may look like a rather academic example of just another nice quirk of nature, having not much practical significance, in reality — it's quite the opposite. Electric power transmission lines, for example, swing in the blowing at a constant speed winds because of these periodically born and released vortices. And this inevitably creates quite powerful at times extra stress at the points of wire fastenings to the supporting towers, which, if neglected, could (and, unfortunately, have) lead to their breakage (sometimes very dangerous). The same holds for, say, tall industrial stacks.

Yet, probably the most renowned engineering calamity of this kind happened in Tacoma (Washington), in November 1940, when a new auto bridge (two lanes, about half a mile long) built just a couple of months before, Fig. 16.4, had started swaying and swinging violently and shortly had collapsed, luckily not killing or injuring anybody (except for the legendary dog mentioned in the epigraph).

A special committee from the Federal Works Agency had been appointed to investigate the accident, with T. von Karman as one of its members, by the way. And the committee's conclusion read that the Tacoma Narrows Bridge had crashed due to "forced vibrations excited by random action of turbulent wind". Well, a while after, a new bridge was built, but this

[a]T. von Karman, (1881–1963), Having greatly contributed to the field of fluid dynamics, he is known as the father of supersonic flight. Hungarian born, he worked for the US government during the World War II, and then for NATO Aeronautical Research and Development Advisory Group. Chairman of Belgium Institute for Fluid Dynamics (now called Von Karman Institute, VKI).

Fig. 16.4: Growing oscillations excited by turbulent vortices led to the collapse of Tacoma bridge.

time with completely different profile of the wind-exposed surfaces, having eliminated, thereby, the cause for unruly vibrations.

Chapter 17

The footprints on the sand

Have you ever thought if walking on a beach we compress the sand under our feet? On the face of it stepping onto the sand one presses the grains together and rams it. But actually the things may turn quite the opposite way. Here is the proof — the footprints left on the wet sand of a sea or river beach stay dry quite a while. The English scientist known for his works on hydrodynamics, O. Reynolds[a], noted in his talk at the meeting of the British association in 1885, that when the foot had stepped onto the sand still wet after the ebb, the surrounding area immediately became dry... According to him the pressure of the foot loosens the sand and the stronger it has been the more water is absorbed. This makes the sand dry until enough water comes from below.

But why does pressure widen intergranular spaces so that the available water suffice no longer to fill them? For scientists of nineteenth century this was not an idle question. The answer was immediately related to the atomic structure of matter. This makes the topic of the present chapter.

17.1 The dense packing of balls

Is it possible to fill the entire space with rigid balls of the same radius? Of course, not, there will always be voids in between[b]. The fraction of space

[a] O. Reynolds, (1842–1912), English physicist and engineer, specialist in theory of turbulence, theory of viscous flows and theory of lubrication.

[b] In principle, it is possible to fill the space with balls provided that their radii $r_1, r_2, \ldots$ form an infinite sequence and $\lim_{n\to\infty} r_n = 0$. However this is of little interest for solid-state physics. —*A. A.*

occupied by the balls is called the ***packing density.*** The closer the balls lie, the less space is left between them, the higher is the packing density. But when does the density reach the maximum? The answer to this question gives the clue to "the mystery of the footprints on the beach".

Let us start from the simpler case and study packings of equal circles in the plane. Dense packings of circles can be obtained by inscribing those into the cells of mosaics *(tilings of the plane)* composed of equal regular polygons. There are only three variants: to cover the plane by equilateral triangles, by squares and by hexagons. The packings of circles in the cells of the square and hexagonal mosaics are shown in Fig. 17.1. It is seen by eye that the second pattern *(b)* is more economical. The accurate calculation (that you may carry out yourselves) proves that in this case 90.7 % of the surface is covered by circles whereas for squares, *(a),* the portion makes only about 78 %. The hexagonal packing is the most dense possible in the plane (or, as modern physicists like to say, in two dimensions). Maybe this was the reason for the bees to use it for the honeycomb.

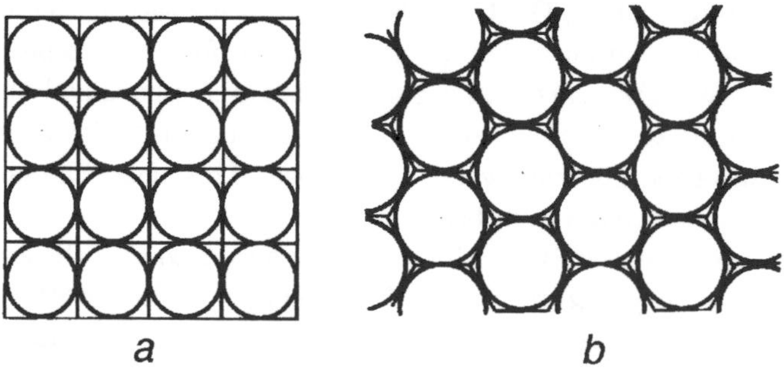

Fig. 17.1: Planar packings of circles inscribed into the cells of tilings of the plane by equal regular polygons.

A dense spatial packing of balls may be realized as follows. Let us first prepare a dense plane layer of balls placed on a flat surface in the economical order described above. We shall call it the X-layer. Now try to put onto it the similar hexagonal second layer. We could do this so that each of the balls of the upper layer would lie right over a ball of the lower one, as if we were filling cells of invisible honeycomb. However in this XX-packing too much empty space will be left. The volume occupied by balls when laid in this manner is only 52 % of the whole.

It is clear how to augment the density. One must simply put upper balls into the holes formed by three touching balls below. (This may be called the XY-arrangement). But it is impossible to fill all the holes at once — one of the two adjacent holes will always remain free, Fig. 17.2. Therefore when putting on the third layer we face a choice. We may either put balls above the holes of the ground X-layer which were left free by the second Y-one (one of these points is denoted by A in Fig. 17.2, *b*) and build a new Z-layer; or we can place them right over the balls of the base (the point B in Fig. 17.2, *b*) in the X-order. Regular spatial patterns are obtained if successive layers follow periodically one of these prescriptions, that is, either $XYZXYZ\ldots$ or $XYXY\ldots$ As a result we obtain the two ways of spatial packing of balls depicted in Fig. 17.3. In both cases the balls fill about 74 % of the space.

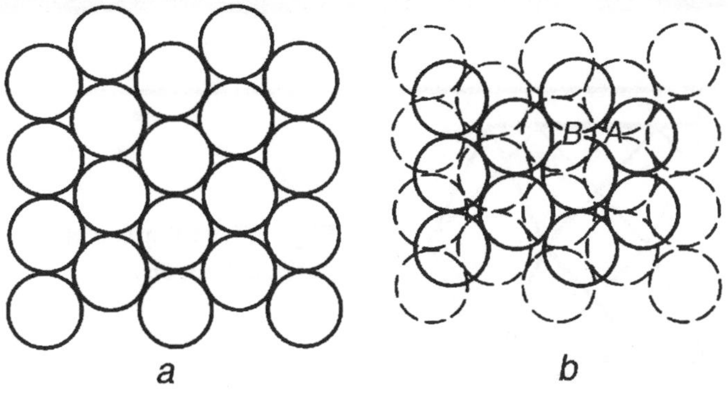

Fig. 17.2: Compact spatial packings of equal balls. (Dashed lines show the lower layer.)

It is easy to count that every ball in these packings touches 12 others and the points of contact are vertices of a 14-faced polyhedron[c]. The faces of

[c]The Greek word *tettarakaidecahedron* (τετταρακαιδεκαεδρον) is out of common use. —A. A.

these polyhedra are alternating squares and equilateral triangles. Say, the first choice (Fig. 17.3, *b*) produces the cuboctahedron[d] shown in Fig. 17.4.

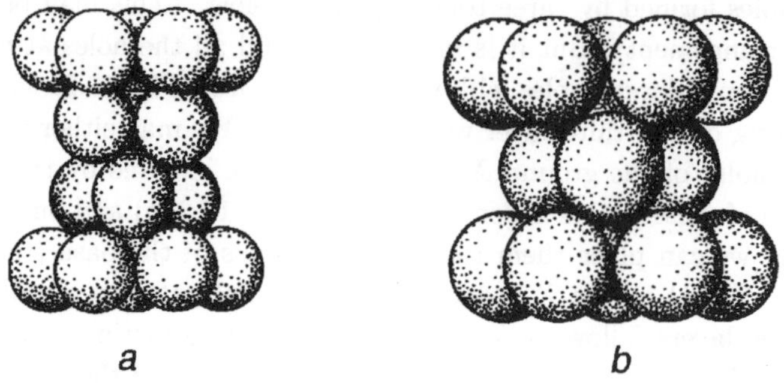

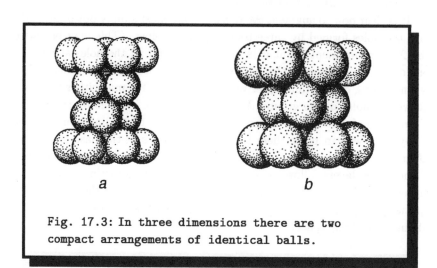

Fig. 17.3: In three dimensions there are two compact arrangements of identical balls.

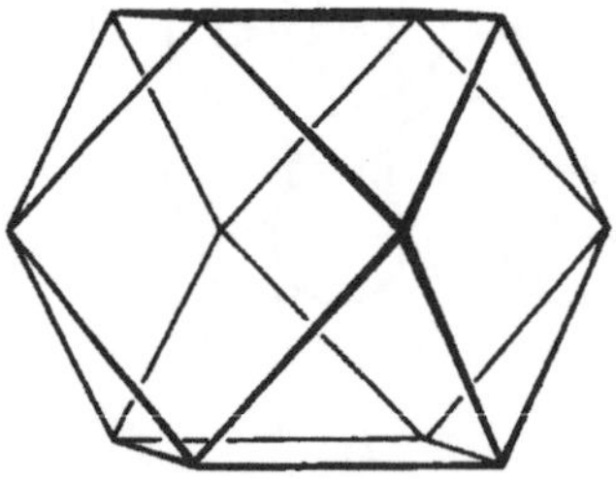

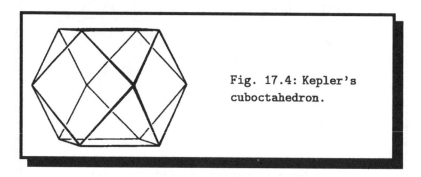

Fig. 17.4: Kepler's cuboctahedron.

So long we only studied how to arrange balls so that they fitted a periodical spatial honeycomb[e]. But is it possible to construct a dense packing

[d]Cuboctahedron belongs to the so-called Archimedes solids. This class comprises 13 convex polyhedra with congruent vertices and surfaces composed of regular polygons of two different types. The name *cuboctahedron* itself belongs to J. Kepler (see page 155). The packing in Fig. 17.3, *b* does not correspond to any of Archimedes bodies since it generates two different types of vertices. —*A. A.*

[e]Another way to say this is that the centers of the balls formed a periodic lattice.—*A. A.*

without this condition? An example is presented in Fig. 17.5. The balls in planes mark out the sides of concentric regular pentagons. The nearest balls of the same pentagon are in contact, but pentagons within a layer are separated. The sides of pentagons in alternating layers contain even and odd numbers of balls respectively. The packing density of this arrangement is 72 % that is not much less than that of the dense hexagonal arrangements shown in Fig. 17.3. There is a way to pack balls so that the centers do not form a lattice whereas the packing density reaches even 74 %. Yet the question whether denser packings exist remains open up till now.

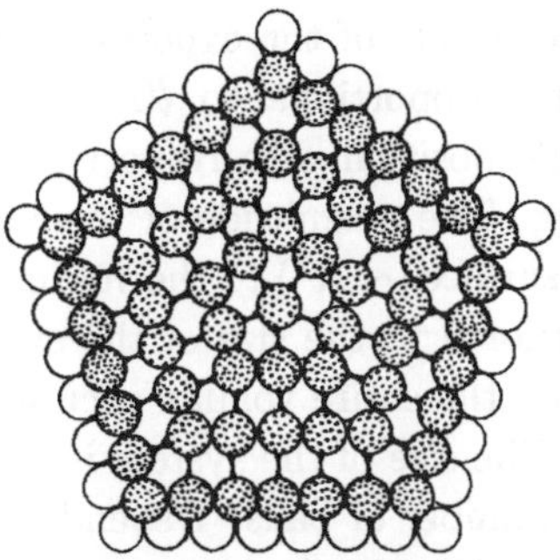

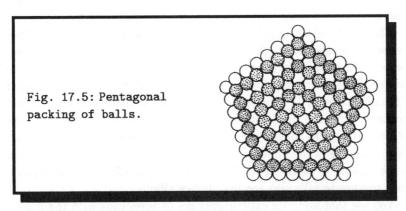

Fig. 17.5: Pentagonal packing of balls.

Let us return to the footprints on the beach. Now we know that there are special arrangements of balls that leave only a small amount of voids in between. If one disturbs such an arrangement displacing, for instance, the balls of one layer away from the holes of the next one the voids will grow. Certainly nobody has cared about arranging the grains of sand in a special way. But how could we force sand to be densely packed?

Remember the common wisdom. What do you do when pouring grain into a can? You shake or pat the can in order to fill it better. Even if the grain has been rammed patting helps to fit something in.

The scientific investigation of this trick was undertaken in fifties by the British scientist G. Scott who was loading spherical flasks of different radii by ball bearings. When they were filled without shaking, so that the balls found places by occasion, the empirical dependence of packing density on the number of balls had the form:

$$\rho_1 = 0.6 - \frac{0.37}{\sqrt[3]{N}},$$

where N was the number of balls. You can see that in case that the number of balls is very big (it has reached several thousands in the experiments) then the density tends to become constant and corresponds to the filling of 60 % of space. But shaking the container in process of being filled helps to increase the density:

$$\rho_2 = 0.64 - \frac{0.33}{\sqrt[3]{N}}.$$

Yet even in this case it is much less than 74 % which corresponded to the regular packing of balls.

The results of the experiment deserve attention. Why is the addition inversely proportional to $\sqrt[3]{N}$? The balls near the walls of the flask are in a special position with respect to those inside. This affects the density of packing. The magnitude of their contribution is proportional to the ratio of the surface ($\sim R^2$) to the volume ($\sim R^3$) of the container being inversely proportional to the size of the system (R). By the volume of the system we mean the entire volume occupied by the balls including spaces between them. The size of the system is $R \sim \sqrt[3]{N}$ since the volume is proportional to the number of balls. Dependencies of this type often appear in physics when one has to take into account surface effects.

You see that accurate experiments are in agreement with the common sense and prove that shaking of granular substances helps to enlarge the density of packing. But still, what is the reason? Remember that positions of stable equilibrium always correspond to minima of potential energy. A ball may forever lie steadily in a hole but it will immediately roll down from a bump. Something of the sort happens here as well. Shaking the flask makes balls roll to free spaces so that the density of packing is increased and the total volume of the system becomes smaller. As a result the level of balls in container goes down. Consequently, the center of mass of the system gets lower and the potential energy is decreased.

Now, at last, we can figure clearly enough what happens to the wet sand. Incessant surf agitates it until a dense packing of grains is formed. Stepping onto the sand by foot we disturb the arrangement of granules and augment intergranular pores[f]. Water from the upper sand layer soaks down to fill the pores. This looks like "drying" of the sand. Taking the foot away

[f]Note that according to Reynolds (see page 149) this refers to the sand *around* the footprint whereas right under the sole it stays densely packed. —*A. A.*

restores the dense packing and the depression left by the foot gets filled with water expelled from narrowing voids. It may happen though that after strong compression the dense packing can not be recovered. Then the footprint will become wet only as water rises from below and fills the widened pores.

It is interesting that these features of granular matters were well known to Indian fakirs. One of their tricks consisted in sticking several times a long thin dagger into a narrow-necked vase poured with rice. At some moment the dagger got stuck in the rice and it was possible to lift the vase holding it by the dagger handle.

Evidently the secret was that piercing randomly poured rice helped to "optimize" the packing of grains just like the shaking would. One may imagine this like a sort of compression wave propagating in a loose medium. In the beginning the grain was packed compactly right around the blade but lay freely in the bulk and near the walls. The "front" of the wave (sure a rather smooth one) divided the dense core from the loose environment. The front had been advancing further with every next stab and when at last it reached the walls of the vessel the rice throughout the volume became densely packed. In other words the possibilities to compress it further were exhausted. The properties of the substance changed drastically, it became "incompressible". And this was the moment when the dagger got stuck, since the pressure of grains onto the blade and consequently the friction were enough to hold it[g].

Caution! In case that you've decided to surprise your party fellows, please, avoid taking glass flasks and china vases. The result can be quite unexpected.

17.2 The long-range and short-range orders

Of course the atoms of which all the bodies are built are far not the rigid balls. Yet simple geometrical arguments help to understand the structure of matter.

The first one to apply geometrical approach was the German scientist Johann Kepler[h] who in 1611 put forward the idea that the hexagonal form

[g]The paragraph was added to the English edition by —*A. A.*

[h]J. Kepler, (1571–1630), German astronomer, creator of celestial mechanics. The famous

of snowflakes is related to the dense packing of balls. Mikhail Lomonosov[i] in 1760 first delineated of the most compact cubic packing of balls and used that to explain the forms of crystal polyhedra. The French abbot R.-J. Haüy[j] noticed in 1783 that all crystals may be constructed of a plethora of repeated parts, Fig. 17.6. He explained the regular form of crystals by suggesting that they are built of identical little "bricks". Finally in 1824 the German scientist A. l. Seeber proposed the model of crystal composed of regularly set little spheres interacting like atoms. The dense packing of the spheres corresponded to the minimum of potential energy.

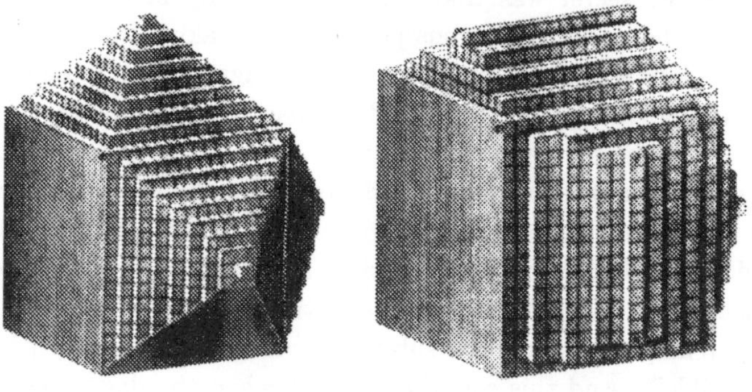

Fig. 17.6: Pictures from the atlas by R.-J. Haüy that was published in the beginning of XIX-th century.

Structures of crystals are the subject matter of the special science called ***crystallography***. Presently periodic arrangement of atoms in crystals is a well-established fact. Electronic microscopes offer a possibility to see this

Kepler laws of planetary motion laid the base for Newton's discovery of the law of gravitation. Kepler's interest in polyhedra was a tribute to the idea of the world ruled by mathematical harmony. According to Kepler ratios of the radii of planetary orbits in the Solar system could be related to properties of regular and uniform polyhedra.

[i]M. V. Lomonosov, (1711–1765), the first scientist of the world importance in Russian history. Successfully worked in natural sciences including physics, chemistry, material science as well as in literature, poetry and painting. Founded the Moscow University.

[j]R.-J. Haüy, (1743–1822), French crystallographer and mineralogist.

by eye. The tendency toward close packing positively does exist in the atomic world. About 35 chemical elements crystallize so that their atoms are situated like the balls depicted in Fig. 17.3. Centers of atoms (or, to be precise, atomic nuclei) make up in space a so-called ***crystal lattice*** that consists of periodically repeated units. The elementary lattices that may be constructed by shifting periodically just a single atom are called ***Bravais lattices*** (after the French naval officer Auguste Bravais[k] who was the first to develop the theory of spatial lattices).

There are not so many Bravais lattices, namely only fourteen of them. The reason is that far not every symmetry element survives in periodic lattices. For instance, a regular pentagon may be turned around the axis passing through the center and it will coincide with the original five times per revolution. One says that it has a fivefold symmetry axis. However a Bravais lattice can not have a fivefold axis. If such a lattice existed the nodes of must be vertices of regular pentagons and those, in turn, would cover without breaks the entire plane. But it is well known that there is no tiling of plane by pentagons, Fig. 17.7!

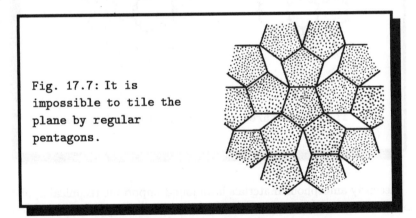

Fig. 17.7: It is impossible to tile the plane by regular pentagons.

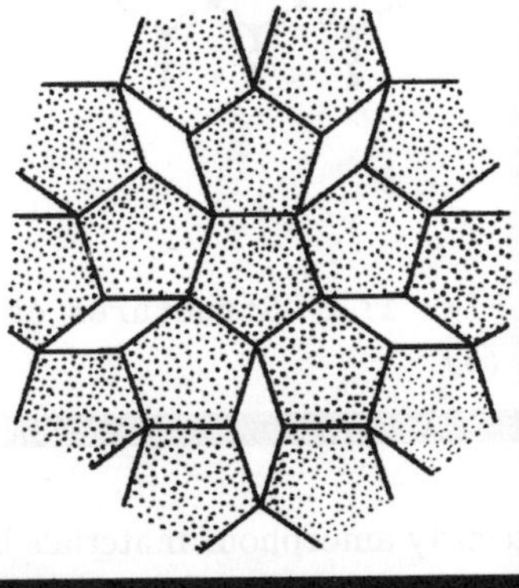

So all crystals can be composed of repeated units. This property is called ***translational symmetry***. One may also say that there is a ***long-range order*** in crystals. Probably this is the main property that distinguishes crystals from other bodies.

There is, though, a none less important class of substances which are

[k]A. Bravais, (1811–1863), French crystallographer.

deprived of the long-range order. These are *amorphous* substances. An example of amorphous state is presented by liquids. But solid matter also may be amorphous. The structure of glass is portrayed in Fig. 17.8 together with that of quartz which has the same chemical composition. But quartz is a crystalline substance in distinction from the amorphous glass. The absence of a long-range order does not mean that atoms in glass are situated chaotically. You may see in the picture that certain ordering of the nearest neighbors is preserved even in glass. This is called a *short-range order*.

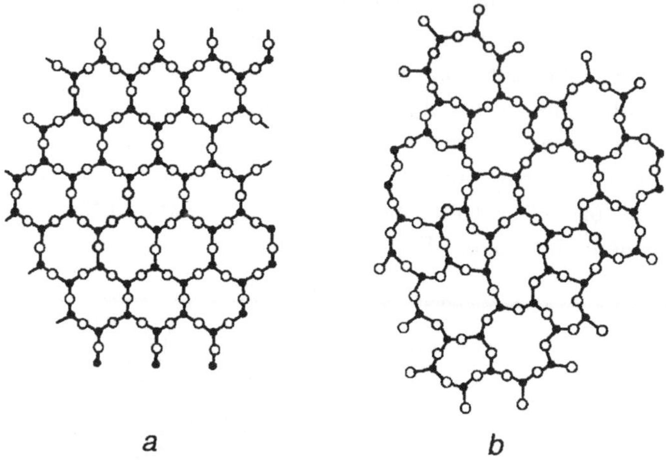

Fig. 17.8: Structures of quartz, *(a)*, and glass, *(b)*.

Recently amorphous materials have found important technical applications. For example, amorphous metal alloys *(metallic glasses)* are marked by unique properties. It turned out that they possess enhanced hardness, high corrosion resistance, exhibit an optimal compromise of electric and magnetic characteristics. Metallic glasses are obtained by means of extremely fast cooling of liquid metal: velocity of cooling must be of the order of several thousands degrees per second. This may be realized by sputtering tiny droplets of metal onto the surface of rapidly spinning cold disk. Droplets get squashed against the disk forming a film of several microns thickness and the instantaneous removal of heat simply leaves no time for the atoms to arrange properly when cooling[l].

Interesting studies that shed light on the structure of amorphous solids were carries out in 1959 by the English scientist J. Bernal[m]. Equally-sized balls of plasticine were randomly put together and compressed into a big lump. The polyhedra obtained after disjoining them back turned to have mainly pentagonal faces. The experiment was repeated with lead pellets. If the pellets had been laid densely and regularly then the compression remolded them into almost regular rhombododecahedrons[n]. On the other hand, if pellets had been poured unintentionally they transformed into irregular 14-faced polyhedra. Among the faces of those were tetragons, pentagons and hexagons forms, but pentagons prevailed.

In modern technology it is often necessary to densely pack the elements of a contrivance. For instance, Fig. 17.9 shows cross section of a superconducting cable made of large number of superconducting wires enclosed in a copper envelope. At first the wires had been cylindrical but after rolling they became hexagonal prisms. The more densely and accurately have been the wires packed the more regular hexagons are seen at section. This is an evidence of the high quality of the cable. If density of packing is interrupted pentagons appear at section.

Fivefold symmetry is widespread in nature. Figure 17.10 presents a photograph of a viral colony. What a striking similarity it has with the pentagonal packing of balls portrayed in Fig. 17.5! Paleontologists even use the presence of fivefold axes in fossils as a proof of their biological (contrary to geological) origin... See what far away from the deserted beach the footprints have led.

[l]In the method of melt spinning, a jet of molten metal is propelled against the moving surface of a cold, rotating copper drum. A solid film of metallic glass is spun off as a continuous ribbon at a speed that can exceed a kilometer per minute.

[m]J. Bernal, (1901–1971), English physicist, specialist in X-ray diffraction analysis, studied structures of metals, proteins, viruses *etc.*

[n]Rhombododecahedron (or rhombic dodecahedron) is a polyhedron with twelve rhombic faces and 14 vertices. It may be obtained as a result of uniform squeezing of the hexagonal packing depicted in Fig. 17.3, *a*. A way to construct it geometrically is to draw mutual tangential planes of a ball and all its neighbors. In the compact packing shown in Fig. 17.3, *a* it plays the role of the unit cell which belongs to hexagons in the planar tiling, Fig. 17.1, *b*. Each ball is inscribed into a rhombododecahedron and those fill the entire space. —*A. A.*

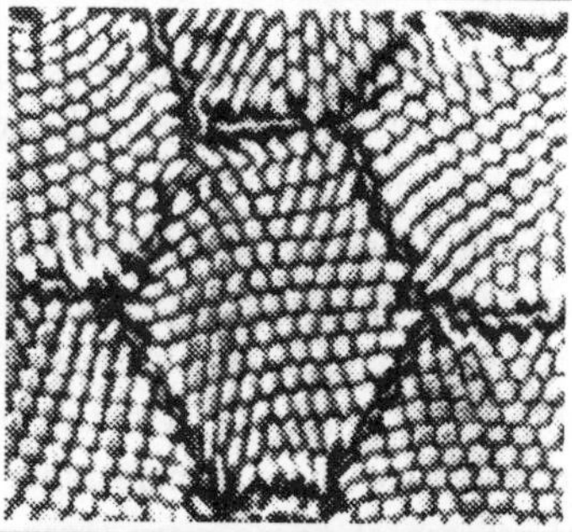

Fig. 17.9: Cross section of high-quality superconducting cable. After rolling cylindrical wires became hexagonal.

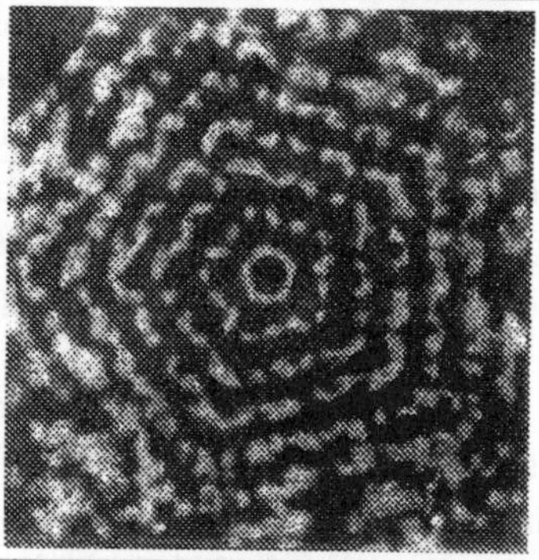

Fig. 17.10: Electron-microscope photograph reveals the fivefold symmetry of viral colony.

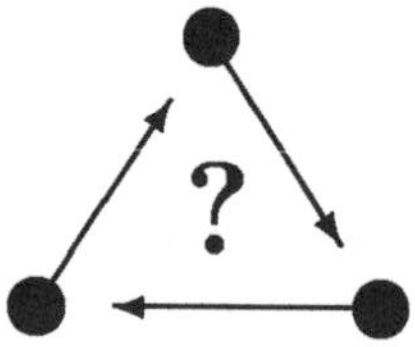

Was it really important that Indian fakirs took for their trick metal vases with long slim necks? What must be the ratio of volumes of the neck and the body of the vase?

Chapter 18

How to prevent snowdrifts

Sections of roads and railroad tracks passing through hollows are often covered with snow even if there have been no snowfalls recently. Why does this happen? Certainly, the answer is at the surface: the snow has been brought by wind. However it took a good deal of investigation to understand in detail the mechanism of the process.

In 1936 the English geologist Ralph Bagnold[a] studied wind transport of sand in air tunnel. He has discovered that unless the wind velocity is larger than some critical value v_1 the sand does not move. If the velocity of air flow is higher than v_1 but yet less than another value v_2 the mass of sand may stay at rest. However an occasional sand grain falling from above brings on several rebound particles. These ones get caught by the wind and, when falling down, knock out more sand grains from the base layer. As a result the sand gets carried by the wind in a kind of leaping motion. In case that velocity exceeds v_2 the wind lifts and blows along substantial clouds of sand. The density of the clouds decreases with height though. Trajectories of sand grains may be viewed in Fig. 18.1.

Now we can explain why wind fills dips with snow. Look at the picture of flow lines in Fig. 18.2. It is obvious that when traversing a hollow the air flow widens and its velocity lessens. This disturbs the balance between deposited and lifted particles: more particles fall down than are taken away. So the depression gets gradually filled with snow.

Analogous processes take place when the snow carried by wind encounters an obstacle, say, a tree. Meeting the trunk the incident flow turns up

[a]R. A. Bagnold, (1896–1990), English geologist, expert in the mechanics of sediment transport and eolian (wind-effect) processes.

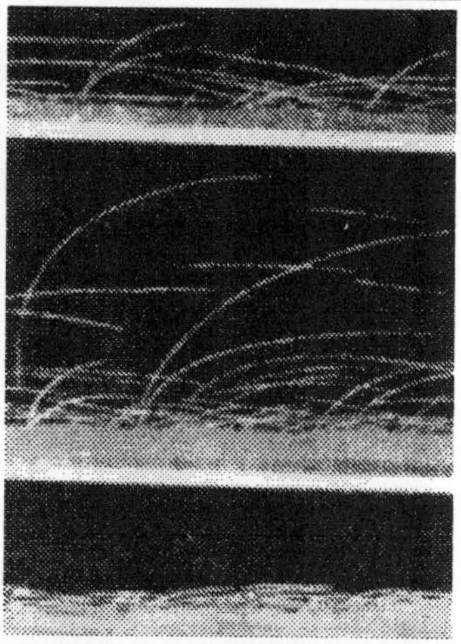

Fig. 18.1: Trajectories of sand particles in air flows of different strength.

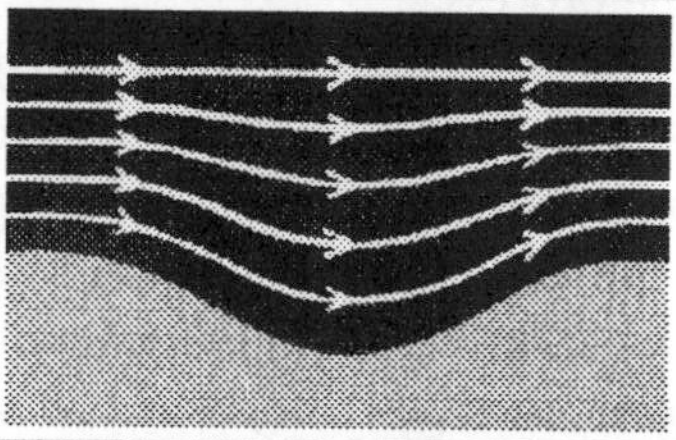

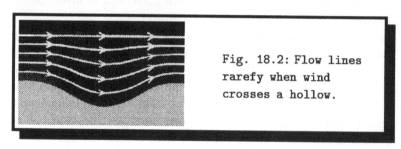

Fig. 18.2: Flow lines rarefy when wind crosses a hollow.

and the ascending current of air is formed. This current digs a deep hole on the windward side of the tree. In the same time in front of the hole and behind of the tree the speed of wind is smaller and a mound is heaped up.

This phenomenon helps prevent snow-binding of low-lying road sections. A protecting fence is made of wooden planks at a certain distance windward of the road. Behind the fence a lee zone of steady light wind is established where all the snow precipitates.

The same mechanism explains the motion of sand dunes. A strong enough wind blowing against a dune picks up sand on the windward side. On the rear of the dune the air flow slows down and the sand falls back down. As a result, with time dunes get inch-by-inch moved along with the wind, the dunes "wander".

Chapter 19

The incident in the train

Not so long ago the authors of these lines had to return from Venice to Naples by express train. The train ran fast (its velocity was about $150\,km/h$) and landscapes that looked like paintings by Renaissance masters flitted by outside the window. In perfect accord with their canvases, the country was hilly, and we sometimes flew over a bridge or dove into a tunnel.

In one of especially long tunnels between Bologna and Florence, we suddenly felt a dull pain in our ears, as happens to air passengers when taking off or landing. It was clear that the same sensation visited all our fellow travelers, who swayed their heads trying to get rid of the unpleasant feeling.

But when the train finally burst out from the narrow tunnel the discomfort passed, however one of us, who wasn't used to such surprises on the railways, got interested in the origin of this phenomenon. Since it was evidently connected with a pressure jump, we started a lively discussion of possible physical causes.

At first glance it seemed to us that the air pressure in the gap between the tunnel walls and the train had increased in comparison with the atmospheric one, but this assumption became less and less obvious as the discussion went on. In such matters mathematics is the best judge, so we attempted to approach the problem quantitatively. Soon the explanation was ready and it came down to this.

Let's consider a train with a cross-sectional area S_t that moves at a velocity v_t in a long tunnel with a cross-sectional area S_0. First of all, let's switch to the inertial coordinate frame associated with the train. We'll take the air flow as stationary and laminar and ignore its viscosity. The motion of the tunnel walls relative to the train need not be taken into account in this case — because of the absence of viscosity it doesn't influence the air flow. We'll also assume that the train is sufficiently long so that one could ignore end effects near the front and rear cars. We shall assume that the air pressure in the tunnel is constant and does not vary along the whole train.

You see, that gradually eliminating minor details, we'd passed from the actual movement of the train to a simplified physical model that could be analyzed mathematically. Here it is.

We have a long tube (the former tunnel) with air being blown through it and a cylinder with streamlined ends (the former train) coaxially fixed inside[a], Fig. 19.1. Far away from the train (at the cross section $A-A$) the air pressure equals the atmospheric one p_0. Velocity of the air flow at this section is equal in magnitude and opposite in direction to the velocity of the train $\vec{v_t}$ with respect to the ground. Let's examine some cross section $B-B$ (just in case, we may place $B-B$ far enough from the ends of the train so that our assumptions were justified). We'll denote the air pressure in this cross section by p and the air velocity by v. These values can be linked with v_t, and p_0 by means of the Bernoulli[b] equation:

$$p+\frac{\rho v^2}{2}=p_0+\frac{\rho v_t^2}{2}, \qquad (19.1)$$

where ρ is the air density. Equation (19.1) contains two unknowns, p, and v; so in order to determine p we need one more relation. This is provided by the condition of conservation of the air flow. According to it the mass of air passing through any cross-section of the tube in a unit of time is constant and equals:

$$\rho v_t S_0 = \rho v (S_0 - S_t). \qquad (19.2)$$

This equation expresses the fact that the air mass can neither appear nor disappear while it flows through the tube. It's usually called the condition of flow continuity.

[a]Note that actually we replaced the ordinary railway tunnel by an air tunnel like those where airplanes are tested.

[b]Daniel Bernoulli, (1700–1782), Swiss physicist and mathematician born in the Netherlands (son of the Swiss mathematician Johann Bernoulli); formulated the fundamentals of theoretical hydrodynamics.

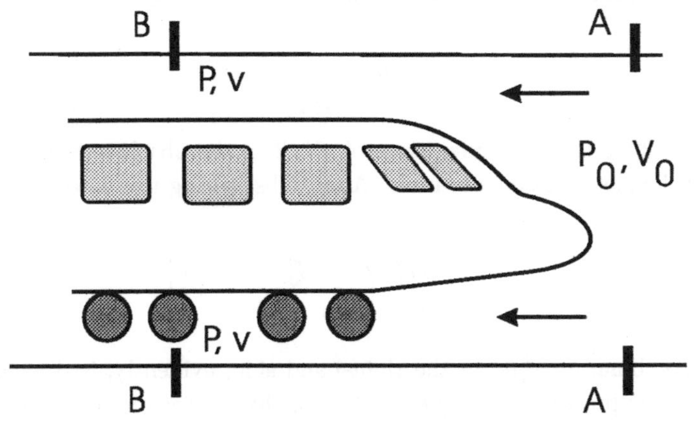

Fig. 19.1: Air flow around fast train in a tunnel is very similar to that in the air tube.

As you probably have noticed, we take the air density in equations (19.1) and (19.2) to be constant. For this assumption to be valid two conditions must hold. The first one requires that the pressure jump we are looking for, Δp, must be much less than the pressure itself: $\Delta p \ll p$. If the air temperature does not change along the tube, then, according to equation (19.4), its density is proportional to pressure: $\rho \propto p$. For small Δp we may neglect the density change $\Delta \rho = \rho \frac{\Delta p}{p} \ll \rho$. We shall see later that this is indeed the fact. The second condition concerns flow velocities at different sections of the tunnel. In order that the density was uniform throughout the tube there must be enough time for air to come to equilibrium. This means that velocities of the flow must be much less than the root-mean-square velocity of chaotic thermal motion of the molecules. It's just this velocity that determines the characteristic time required to establish the constant equilibrium gas density on the macroscopic scale.

Eliminating the velocity v from equation (19.1) by means of equation (19.2), we get:

$$p = p_0 - \frac{\rho v_t^2}{2}\left(\frac{S_0^2}{(S_0 - S_t)^2} - 1\right). \qquad (19.3)$$

The air density ρ can be expressed in terms of p by the Mendeleyev-

Clapeyron equation (as we have already done in Chapter 12, see equation (12.3)):

$$\rho = \frac{m}{V} = \frac{p_0\,\mu}{RT}, \tag{19.4}$$

where $\mu = 29$ is the molecular mass of air, T is the absolute temperature and R is the gas constant per mole. After substituting this into (19.3), we get:

$$p = p_0\left[1 - \frac{\mu\,v_t^2}{2RT}\left(\frac{S_0^2}{(S_0 - S_t)^2} - 1\right)\right]. \tag{19.5}$$

The combination $\frac{\mu\,v_t^2}{2RT}$ in the right-hand side, evidently, is dimensionless. So the expression $\sqrt{RT\,/\,\mu}$ must have the dimension of velocity. Up to a coefficient it's easy to recognize in it the root-mean-square velocity of thermal molecular motion. But in aerodynamical problems another physical characteristic of gas, the sound velocity v_s, is more to the point. It is determined by the same combination of temperature and molecular mass as the root-mean-square velocity of molecules, but the numerical value of v_s includes in addition the so-called adiabatic index γ. The latter is a number of the order of unity characteristic of a gas (for air, $\gamma = 1.41$):

$$v_s = \sqrt{\gamma\frac{RT}{\mu}}\,. \tag{19.6}$$

Under normal conditions, $v_s = 1\,200\,km/h$. With the help of equation (19.6) we can bring the expression (19.5) to the form that will be convenient for the further discussion:

$$p = p_0\left[1 - \frac{\gamma}{2}\frac{v_t^2}{v_s^2}\left(\frac{S_0^2}{(S_0 - S_t)^2} - 1\right)\right]. \tag{19.7}$$

Now it's time to stop and think a little. We calculated the pressure along the skin of the train inside the tunnel. But our ears ached not because of the pressure itself but because it had changed in comparison with the pressure p' that was there when moving in the open[c]. We can easily determine this

[c]Here we should point out two circumstances. First, in biophysics there is the so-called Weber-Fechner law. According to it any changes in the environment can be detected by organs of sense only if the relative change of parameters exceeds some threshold value. The second is that in a long tunnel our organism adapts to the new conditions and the discomfort disappears. Nevertheless, it comes back at the exit.

pressure directly from equation (19.7), noticing that the open air can be considered as a tunnel of the infinite cross-section $S_0 \to \infty$. So we have:

$$p' = p_0.$$

This result was sufficiently evident without calculation, though.

It's interesting to observe that the relative pressure difference is negative:

$$\frac{\Delta p}{p_0} = \frac{p - p_0}{p_0} = -\frac{\gamma}{2}\left(\frac{v_t}{v_s}\right)^2 \left(\frac{S_0^2}{(S_0 - S_t)^2} - 1\right). \tag{19.8}$$

From this we can see that when a train enters a tunnel the pressure around it decreases, contrary to what we might expect at first. Now let's estimate the magnitude of the effect. As we have mentioned before, $v_t = 150\, km/h$ and $v_s = 1\,200\, km/h$. For narrow railroad tunnels the ratio $S_t \,/\, S_0$ is about $1/4$ (for there were two tracks in our tunnel). So we find that:

$$\frac{\Delta p}{p_0} = -\frac{1.41}{2}\left(\frac{1}{8}\right)^2 \left(\left(\frac{4}{3}\right)^2 - 1\right) \approx 1\,\%.$$

This value may seem pretty small, but if we take into account that $p_0 = 10^5\, N \,/\, m^2$ and take the area of the eardrum to be $\sigma \sim 1\, cm^2$, we get an excess force $\Delta F = \Delta p_0 \cdot \sigma \sim 0.1\, N$, which may turn out to be quite noticeable.

So it seemed that the effect was explained, and we could call it quits. But something still worried us about the final equation. Namely, from expression (19.8) it followed that even in the case of an ordinary train moving with normal velocity, so that[d] $\frac{v_t}{v_s} \ll 1$, the value of $|\Delta p|$ might reach and even exceed the normal pressure p_0 in sufficiently narrow tunnels! Clearly, within the framework of our assumptions we were getting the absurd result that the pressure between the walls of narrow tunnel and train became negative!

Wait a minute! Probably we have missed something that restricts validity of our formula... Let's take a closer look at our findings. If $\Delta p \sim p$, then

$$\frac{v_t}{v_s}\left(\frac{S_0}{S_0 - S_t}\right) \sim 1,$$

[d]The omnipresent in aerodynamics ratio of velocities $M = v \,/\, v_s$ is called the Mach number after the Austrian physicist Ernst Mach, (1838–1916).

and, consequently,

$$v_t\, S_0 \sim v_s\, (S_0 - S_t).$$

Comparing the last equation with the continuity equation (19.2), we begin to understand the situation. If Δp becomes of the order of p_0 the velocity of air flowing in the gap between the train and the walls of this narrow tunnel turns out to be of the order of the speed of sound. That is, we can't speak of laminar air flow any longer and the previously smooth flow becomes turbulent.

So the correct condition for the use of equation (19.8) is not merely $v_t < v_s$ but the more rigid one:

$$v_t \ll v_s \left(\frac{S_0 - S_t}{S_0} \right)$$

It's evident that for real trains and tunnels this condition is always met. Nevertheless, our investigation of the limits of applicability of equation (19.8) wasn't just an empty mathematical exercise. Physicist must always understand the limits of validity of his result. Besides there is a quite practical reason to take it seriously. In the last few decades fundamentally new forms of transportation, including high-speed trains, were discussed more and more. One of the projects exploited magnetic cushion produced by a powerful superconducting magnet. Already in the early nineties a prototype *maglev* (an abbreviation for *magnetic levitation*) train in Japan could carry 20 passengers along the $7\,km$ test track at a maximum speed of $516\,km/h$ — that makes almost a half of the sonic speed! The vehicle hovered above metal rails supported by strong magnetic field and resistance to its motion was determined solely by aerodynamic effects.

The next step in developing this transport was the idea of — believe it or not — enclosing the train in a hermetically sealed tube and reducing the aerodynamical factor by pumping the air out! You see how close this problem is to the one that has captivated us. But here physicists and engineers have encountered a much more complex case of $v_t \sim v_s$ and $S_0 - S_t \ll S_0$. Therefore the air flow is far not laminar, and the temperature of the air changes considerably along the train.

Modern science doesn't have ready answers to questions which appear when solving these problems. But even our simple estimate allows, in principle, to understand when new effects come into play and become important.

By the way, here are some more physical questions that might pop up during a train ride.

(1) Why does the noise from a moving train increase considerably when the train enters a tunnel?
(2) Which of the two rails of a track is worn down faster in the Northern Hemisphere? And what about the Southern Hemisphere?

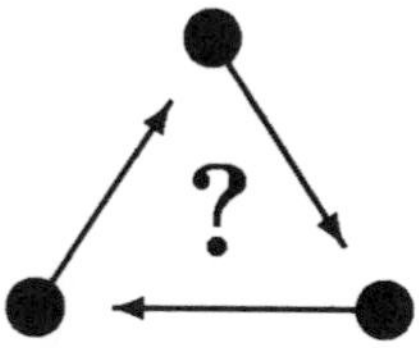

Why do express trains on close parallel tracks slow down when meeting?

PART III

Windows to the quantum world

Finally the time has come to tell you about strange laws ruling in the world of microparticles. You will learn how these laws reveal themselves through "superphenomena" which take place at very low and not so low temperatures. We shall try to make the things as easy as possible. Still this will not be a small talk for any real comprehension would appeal to mathematical language of modern physics. But we hope that you'll appreciate the marvels of this unbelievable, surprising and promising world.

Chapter 20

The uncertainty relation

> Coordinate and momentum resemble the male and female silhouettes in antique barometer. For either of them to show up the second must disappear.
>
> *Werner Heisenberg.*

In 1927 the German physicist Werner Heisenberg[a] discovered the uncertainty relation. Suppose that watching some body we managed to determine the projection of the momentum onto the x-axis with the accuracy Δp_x. Then we shall not be able to measure the corresponding x-coordinate with the precision greater than $\Delta x \approx \hbar \,/\, \Delta p_x$, where $\hbar = 1.054 \cdot 10^{-34}\, J \cdot s$ is the Planck's constant.

At first this relation looks perplexing. Remember the Newton laws that we have studied at school allow to find the equation of motion of a body and to calculate the time-dependencies of the coordinates. Knowing those one can compute the velocity $\vec{v}$ (that is the time derivative of the coordinate $\vec{x}$), the momentum $\vec{p}$ and its projections. It looks as if we had established both coordinates and momenta and there was no uncertainty relation. Indeed, this is the fact in classical physics[b], but the situation changes in the microworld, Fig. 20.1.

[a]W. K. Heisenberg, (1901–1976), one of the foundators of quantum mechanics; Nobel Prize 1932.

[b]Of course the uncertainty relation holds in the macroworld too. But there it is not restrictive and does not play the key role.

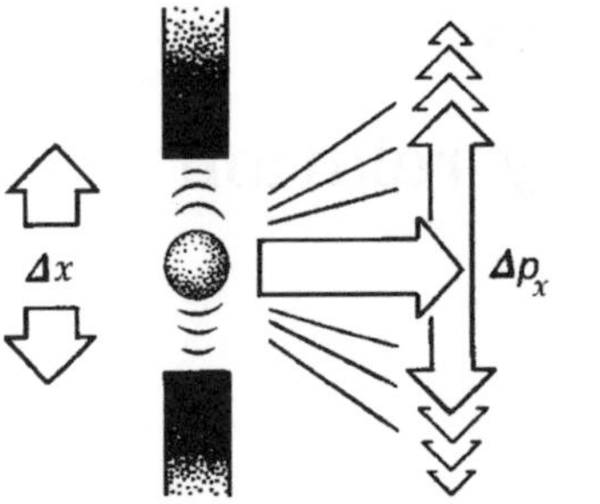

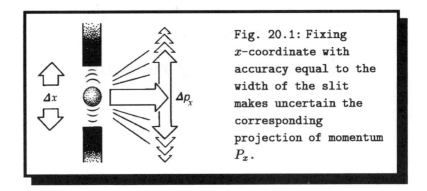

Fig. 20.1: Fixing x-coordinate with accuracy equal to the width of the slit makes uncertain the corresponding projection of momentum P_x.

20.1 Momentum and coordinate

Imagine that we want to trace the motion of an electron. What should we do? Human eyes hardly answer the purpose. Their resolving ability is too weak to discern electron. Well, let us try a microscope. Resolution of a microscope is limited by the wave length of the light used for the observation. Wavelengths of the usual visible light are of the order of $100\,nm$ $(10^{-7}\,m)$ and one can not see smaller particles in a microscope. Sizes of atoms are of the order of $10^{-10}\,m$ and there is no hope to discern them not to mention single electrons.

But let us fantasize. Suppose that we managed to construct a microscope which exploits not visible light but electromagnetic waves of smaller length, say, X-rays or even γ-rays. The harder γ-radiation we use the shorter are the corresponding waves and the smaller objects can be detected. This imaginary γ-microscope seems to be an ideal instrument capable of measuring electron positions with a desired accuracy. And what about the uncertainty relation?

But think over this hypothetical experiment (some physicists like the German word *Gedankenexperiment*) once more. In order to bring the information about the position of the electron at least one γ-quantum must be reflected by it. One quantum carries the minimal amount of energy of the radiation that is equal to $E = h\,\nu = \hbar\,\omega$ where ω is the angular frequency of the field oscillations. Short waves have higher frequencies and their quanta carry bigger energies. But the momentum of a quantum is proportional to the energy. Colliding with an electron a quantum inevitably transfers to the particle a fraction of its momentum. Because of that any coordinate measurement makes the electron momentum ambiguous. The rigorous analysis of the process proves that the product of the uncertainties can not become smaller than the Planck's constant.

You may think that the discussion referred only to this particular case. Maybe we have proposed the "wrong" device and there are more delicate ways to measure the coordinate without "kicking" the electron to a new state. Unfortunately this is not so.

The best scientists (including A. Einstein) tried to invent a device (or a Gedankenexperiment) which could determine position and momentum of a body with accuracy better than what the uncertainty relation prescribed. But all attempts failed. By the law of nature this is impossible[c].

Our arguments may appear vague and there is no self-evident mental model at hand. Real understanding demands serious learning of quantum mechanics. But this was enough to make the first acquaintance with the subject.

In order to mark the boundary between the micro- and macroworlds let us make an estimate. Tiny particles used in observations of Brownian motion are about $1\,\mu m$ ($10^{-6}\,m$) big and weigh less than $10^{-10}\,g$. Still these fragments of matter contain enormous numbers of atoms. The uncertainty relation tells us that for them $\Delta v_x\,\Delta x \sim \hbar\,/\,m \sim 10^{-21}\,m^2\,/\,s$. Suppose that we are going to fix the particle position with the accuracy equal to one percent of the size, $\Delta x \sim 10^{-8}\,m$. Then $\Delta v_x \sim 10^{-13}\,m\,/\,s$. This is a very small quantity and the reason of that is the small value of the Planck's constant.

Brownian velocity of such a particle is approximately $10^{-6}\,m\,/\,s$. Apparently the inaccuracy of the velocity coming from the uncertainty relation is negligible. It is less than one tenmillionth (0.0000001!) even for a body this small. And because of the $\hbar\,/\,m$ in the right side of the relation it does not tell for larger bodies all the more. But if we reduce the mass (take for example an electron) improving in the same time the accuracy of measurements (let $\Delta x \sim 10^{-10}\,m$, the atomic size) the uncertainty of the velocity becomes comparable to the velocity itself. For electrons in atoms the uncertainty relation carries the full weight and may not be ignored. This leads to breath-taking consequences.

[c]Lately attention of physicists was caught by so-called ***squeezed states***. There the product of uncertainties is somewhat smaller but still of the order of $\hbar$. Existence of such exceptional states does not affect the general principle. —*A. A.*

20.2 The probability waves

The simplest model of atom is the Rutherford[d] planetary model where electrons circle the nucleus like planets orbit the Sun. But electrons are charged particles and orbiting brings on varying electric and magnetic fields. This gives rise to electromagnetic radiation which takes away energy. This foretells a bad lot to electrons in the planetary model: after having emitted all the energy they must fall onto the nucleus. Planetary model predicts collapse of atoms. In the mean time stability of atoms is a solid experimental fact.

The Rutherford model required a "touch-up". This was done in 1913 by Niels Bohr[e]. In his model electrons are allowed to occupy only certain orbits with strictly defined energies. Electrons may change the energy only by jumping from one orbit to another. This "quantum" behavior explains many things and among those atomic spectra and stability of atoms. Even now it is helpful in simplified treatment of quantum effects. But it violates the uncertainty relation! Obviously, in contrast to the laws of microworld both coordinate and momentum of electron on orbit are definite, regardless quantum or classical the orbit is.

The further development corrected this so-called semiclassical model. Actual behavior of electrons in atoms proved even more startling.

Suppose that we managed to find where exactly the electron is at the moment[f]. Is it possible to predict positively where it will be a bit later, to be definite, say, in a second? No, because as we know, the coordinate measurement has inevitably introduced the uncertainty into the momentum. Predicting where the electron gets is beyond the power of devices. What should we do?

Let us mark the spatial point where we have found the electron. Another mark will register the result of analogous measurement performed on one more atom of the type. The more measurements, the more marks. It turns out that although it is impossible to tell where the next one appears, the

[d]E. Rutherford, 1st Baron of Nelson, (1871-1937), English physicist; Nobel Prize for chemistry 1908.

[e]Niels H. D. Bohr, (1885–1962), Danish physicist; the first who proposed the idea of quantization; Nobel Prize 1922.

[f]Another problem we meet in quantum mechanics is that electrons are identical and it is impossible to distinguish them. The further discussion implicitly refers to the hydrogen atom which contains only one electron. —*A. A.*

spatial distribution of the marks follows a pattern. The density of marks varies from point to point indicating whether the probability to meet the electron is more or less.

We had to give up the idea to describe motion of electron in detail but still we can judge the chances to locate it at different points in space. Behavior of electron in microworld is characterized by probability! The reader may dislike the strange suggestion. It is absolutely out of habit and contradicts our intuition and routine experience. But there is nothing we can do about the fundamentals of nature. The laws of microworld are really different from those of every day. According to the comparison by A. Einstein we must "cast dice" in order to predict behavior of electrons. One has to face the facts[g].

So, in microworld a state of electron is defined by the probability to find it at various spatial points. In our pictorial model the probability is proportional to the density of marks. One may fancy that the marks form a sort of a cloud where electron lives.

But what regulates the structure of probability clouds? You know that classical mechanics is ruled by Newton laws. Quite similarly quantum mechanics has its own equation which determines "spreading" of electron in space. This equation was found in 1925 the by Austrian physicist Erwin Schrödinger[h]. (Note that this had been before the uncertainty relation cleared the reason of particle spreading. These things happen in physics.) The Schrödinger equation provides exact and detailed quantitative description of atomic effects. But it is impossible to solve it without complicated mathematics. Here we shall cite some ready answers that illustrate electron spreading.

Figure 20.2 presents a scheme of experiment on electron diffraction. The pattern of bands which appears on the screen is shown in the photograph. It is indubitably akin to a pattern of light diffraction. The result would be impossible to explain if electrons followed linear trajectories as prescribed by the laws of classical physics. But if they are smeared in space this is conceivable. Moreover, the experiment demonstrates that probability clouds exhibit wave properties. Probability waves such as birth rate or crime waves are common in day-to-day life. Amplitude of the wave is

[g]We must note that A. Einstein himself did not believe the necessity of such "gambling". He did not accept quantum theory till the end of his life.

[h]E. Schrödinger, (1887–1961), Austrian physicist; Nobel Prize 1933.

maximal at the place of the greatest probability of an event. In our case it is most probable to find the electron there. In the photograph these zones are lighter in color.

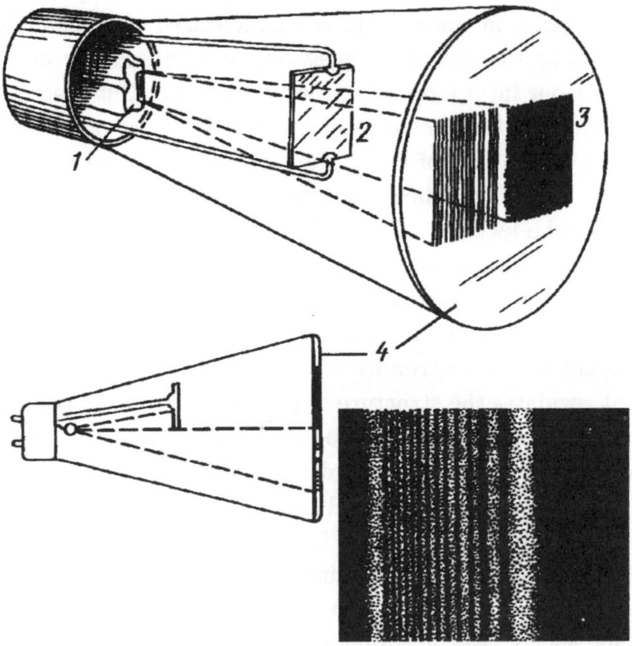

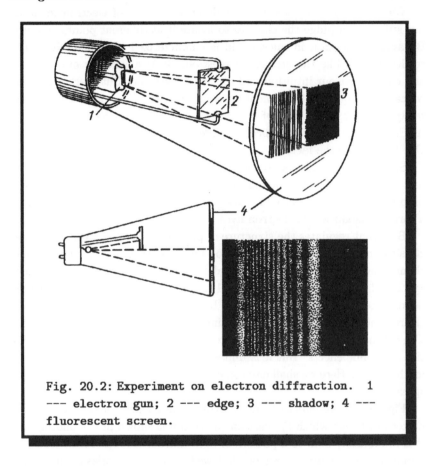

Fig. 20.2: Experiment on electron diffraction. 1 --- electron gun; 2 --- edge; 3 --- shadow; 4 --- fluorescent screen.

Smearing of electron in hydrogen atom as obtained by exact mathematical analysis of several quantum states is portrayed in Fig. 20.3. These are analogues of quantum electron orbits in the Bohr model of atom. Again the probability to meet electron is higher in the lighter regions. The pictures remind snapshots of standing waves in finite domains. Probability clouds are really magnificent! Besides these abstract pictures do indeed define the behavior of electrons in atom and explain, for example, energy levels and all what concerns chemical bonding.

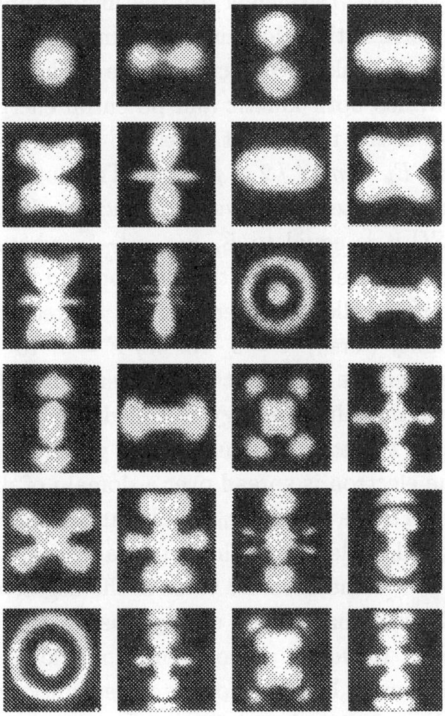

Fig. 20.3: ''Smearing'' of electrons in atom. This is not a photograph but a result of calculations. Symmetries of these pictures strongly influence symmetries of molecules and crystals. One may even say that they are keys to understanding of the beauty of ordered life forms in nature.

The uncertainty relation makes possible to estimate dimensions of probability clouds without going into particulars of their structure. If the size of a cloud is of the order of Δx then it makes no sense to speak of coordinate uncertainties greater than Δx. Consequently, the uncertainty of momentum of the particle, Δp_x, can not be less than $\hbar / \Delta x$. The same expression determines by the order of magnitude the minimal momentum of the particle.

The smaller is a cloud, the bigger are the momentum and velocity of motion within the localization domain. It turns out that these general considerations suffice to make a correct estimate of the atomic size.

Electron in atom has both kinetic and potential energies. The kinetic energy is the energy of the motion. It is related to the momentum by the well-known formula: $E_k = mv^2 / 2 = p^2 / 2m$. The potential energy of the electron is the energy of interaction with the nucleus. In the *International System of Units (SI)* it is equal to $E_p = -\frac{1}{4\pi\varepsilon_0}\frac{e^2}{r}$ where e is the charge of electron, r is the distance between the electron and the nucleus and the dimensional constant is called the permittivity of free space. It may be found from the equality $(4\pi\varepsilon_0)^{-1} = 9 \cdot 10^9\, m / F = 9 \cdot 10^9\, J \cdot m / C^2$.

The electron has definite value of the full energy $E = E_k + E_p$ in every state. The state with minimal energy is called the ***ground (nonexcited)*** state. Let us estimate the atomic radius for the ground state.

Imagine that the electron is spread over a domain of dimension r_0. Attraction to the nucleus tends to make r_0 smaller and collapse the probability cloud. This corresponds to lessening the potential energy that by the order of magnitude is $-\frac{1}{4\pi\varepsilon_0}\frac{e^2}{r_0}$ (the absolute value of the negative quantity grows as r_0 goes down). In case that kinetic energy was not there electron would fall onto the nucleus. However, as you remember, localized particles always possess kinetic energy by virtue of the uncertainty relation. And this prevents electrons from falling down! Decreasing r_0 enhances the minimal momentum of the particle $p_0 \sim \hbar / r_0$ and as a result the kinetic energy $E_k \sim \hbar^2 / 2mr_0^2$ grows too. The full energy of electron E is minimal when the derivative dE / dr_0 is zero. From this we obtain that the minimum corresponds to

$$r_0 \sim \frac{4\pi\varepsilon_0 \hbar^2}{m e^2}. \tag{20.1}$$

This determines the typical dimension of localization domain that is essentially the atomic radius. The obtained value of r_0, (20.1) is $0.05\, nm$ $(5 \cdot 10^{-11}\, m)$. As you know by the order of magnitude this is the actual dimension of atom. Obviously if the uncertainty relation makes possible to correctly estimate atomic radius it must belong to the most profound laws of the microworld.

Another principle that follows directly from the uncertainty relation pertains to complex atoms. ***Ionization energy*** E_i is defined as the work required for detaching an electron from the atom. It can be measured pretty accurately. Imagine that the product of $\sqrt{E_i}$ and the atomic size d is the same for absolutely different atoms up to 10–20%. Probably the reader has already guessed the reason: the momentum of electron is $p \sim \sqrt{2mE}$ and according to the uncertainty relation the product $p \cdot d \sim \hbar$ must be constant.

20.3 The zero-point oscillations

Impressive results come out of applying the uncertainty relation to oscillations of atoms in the solid state. Atoms (or ions) oscillate about the nodes of crystal lattice. Usually the oscillations are due to thermal motion and increase as temperature rises. But what happens if temperature is reduced? From classical point of view the amplitude of oscillations will decrease and atoms will come to halt at absolute zero. But is this possible from the point of view of quantum laws?

Shrinking the amplitude of oscillations means, in quantum language, compressing the probability cloud (or the localization domain) of a particle. We have seen, that because of the uncertainty relation, the price to pay will be the enhancement of the particle momentum. Attempts to arrest a quantum particle fail. It turns out that even at the absolute zero temperature atoms in solids do oscillate. These ***zero-point oscillations*** give rise to a number of beautiful physical effects.

First of all let us try to estimate the energy of zero-point oscillations. In oscillatory system a restoring force $F = -k\,x$ appears as the body is shifted to a small distance x away from the equilibrium. For a spring k is the elasticity coefficient while in solid it is defined by forces of interactions between the atoms. The potential energy of the oscillator is

$$E_p = \frac{k\,x^2}{2} = \frac{m\,\omega^2\,x^2}{2},$$

where $\omega = \sqrt{k\,/\,m}$ is the frequency of oscillations.

This means that the energy of oscillator may be expressed in terms of the amplitude of oscillations $x_{\max}$,

$$E = \frac{m\,\omega^2\,x_{\max}^2}{2}; \qquad x_{\max} = \sqrt{\frac{2E}{m\,\omega^2}}.$$

But in quantum language the amplitude of oscillations is the typical dimension of the localization domain that, because of the uncertainty relation, determines the minimal momentum of the particle. It comes out that, on the one hand, the smaller is the energy of oscillations the smaller the amplitude must be. But, on the other hand, reducing the amplitude increases the momentum and, consequently, the kinetic energy of the particle. The minimal energy of a particle is given by the estimate,

$$E_0 \sim \frac{p_0^2}{2m} \sim \frac{\hbar^2}{m x_0^2} \sim \frac{\hbar^2}{m} \cdot \frac{m\omega^2}{E_0}.$$

Comparison of the last two expressions results into $E_0 \sim \hbar\omega$. The exact calculation gives the twice smaller value. The energy of zero-point oscillations is equal to $\hbar\omega / 2$. It is maximal for light atoms that oscillate with higher frequencies.

Probably the brightest manifestation of zero-point oscillations is the existence of a liquid that does not freeze even at the absolute zero. Obviously a liquid will not freeze if the kinetic energy of atomic oscillations is enough to destroy the lattice. It does not matter whether the kinetic energy appears due to the thermal motion or due to the quantum oscillations. The most likely candidates for nonfreezing liquids are hydrogen and helium. The energy of zero-point oscillations is maximal in these lightest substances. But above that helium is an inert gas. The interaction between helium atoms is very weak and it is comparatively simple to melt the crystal lattice. It proves that the zero-point oscillations energy is enough and helium does not freeze even at the absolute zero. On the contrary the interaction of atoms in hydrogen is much stronger and it freezes despite the zero-point oscillation energy of atoms is greater in hydrogen than in helium.

All other substances do freeze at absolute zero too. So helium is the only one that always remains liquid at normal pressure. One may say that it is the uncertainty relation that prevents it from freezing. Physicists call helium a ***quantum liquid.*** Another exceptional property it has is ***superfluidity,*** which is also a macroscopic quantum phenomenon.

Still under the pressure of about 2.5 *MPa* liquid helium becomes solid. Although the solid helium is not quite an ordinary crystal. For example kinetic energy of atoms at the interface of solid and liquid helium is defined by the zero-point oscillations. This enables the crystal surface to perform gigantic oscillations as if it was an interface of two nonmiscible fluids, Fig. 20.4. Physicists have graced solid helium with the title of ***quantum crystal*** and vigorously investigate its properties.

Fig. 20.4: Waves on the interface between solid (the light region) and liquid helium.

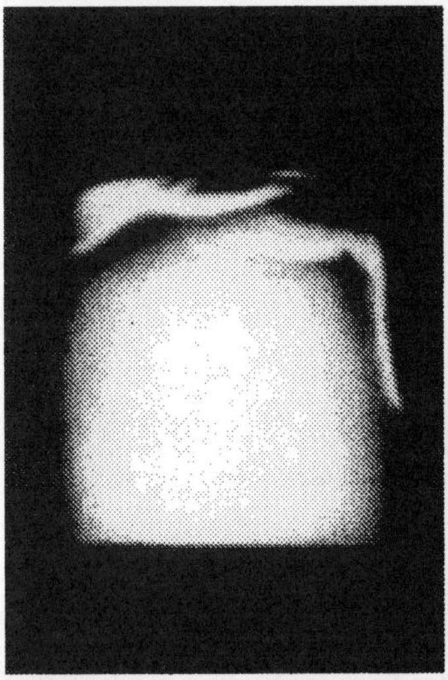

Chapter 21

On the snowballs, nuts, bubbles and... liquid helium

Helium, despite his just second position in the periodic table of elements, has since its discovery been a source of great many hassles for physicists due to its quite unorthodox properties. Yet these troubles and headaches were overwhelmingly outshone by the beauty and uniqueness of physical phenomena occurring in the liquid helium as well as by the opportunities it offers to researchers and engineers in production of extremely low temperatures. Among the peculiarities of this quantum liquid, besides superfluidity, there stands also its specific, different from other liquids, mechanism of charge transfer. And here our story starts.

Physicists started tinkering with this question in the late fifties. At that time the most probable candidates for the role of charge carriers seemed to be electrons and positive ions produced by ionization of helium atoms. The further assumption was that actually electric charge is transported not by the helium ions themselves (they are rather heavy and it would be too hard to accelerate them) but by *the "holes"*. To get an idea of what the "holes" are, you may imagine that an electron sitting in a helium atom has leaped onto a positively charged helium ion which happened nearby. When doing so the electron must have left behind an empty place. However that "electron vacancy" may soon enough be occupied by another electron hopping from another atom. That new empty seat, in turn, will be taken by another electron, and this way it can go on and on. From aside, such an electron "leapfrog-game" looks as if a positively charged particle moved in the opposite direction. Yet since, in fact, there is no real moving positive charge, just an absence of electron at its "dwelling place", we can call this object a "hole". This charge transfer mechanism generally works fine in semiconductors, so it was considered quite plausible that it works in liquid helium.

As a rule, in order to measure masses of charge carriers, both positive and negative, researchers study their trajectories in uniform magnetic field. It's known that when a charged particle with some initial velocity enters magnetic field it starts gyrating and its trajectory becomes a circle or a spiral. Knowing the initial velocity and the strength of the field one can easily find the mass of the particle simply by measuring the radius of gyration. However, the results of experiments turned out to be really surprising: the obtained masses of both negative and positive charge carriers exceeded that of the free electron by tens of thousands of times!

Sure enough, in liquid electrons and holes are surrounded and interact with atoms and, hence, their masses can differ from that of the free electron. Yet five orders of magnitude seemed way too much. Such a significant discrepancy of theoretical calculations and experimental results was considered unacceptable even for the extravagant helium. So there arose an urgent need to come up with a new not known before model.

Shortly, the correct explanation of the structure of the charge carriers in liquid helium was proposed by the American physicist Robert Atkinson. It is known that in order to transform a liquid into solid one doesn't necessarily need to cool it, — it's just as well possible to solidify it by compressing harder and harder. The pressure at which the liquid becomes solid is called the ***solidification pressure*** (P_s). Naturally, P_s depends on temperature: the higher the latter is the more difficult it gets to solidify the liquid by compression, and, therefore, P_s goes up. It turns out that the whole "trick" with the structure of the positive charge carriers is explained by the rather low value of solidification pressure of liquid helium: at low temperatures, $P_s = 25\,atm$. And this causes very unusual structure of the positive carriers.

We've mentioned earlier that positive ions He^+ can commonly exist in the liquid helium. When interacting with a neutral helium atom a positive He^+ attracts the negatively charged electrons and repels the positively charged nucleus. Resulting is that the centers of positive and negative charges in the atom don't coincide anymore but get separated by a distance. Hence presence of positive ions in the liquid helium should lead to ***polarization*** of its atoms. The polarized atoms are attracted by the positive ion and this, in turn, causes a rise of the local concentration of He atoms and the local density. As a result, the pressure around the ion increases. Graphically the dependence of the incremental pressure on distance from positive ion is depicted in Fig. 21.1.

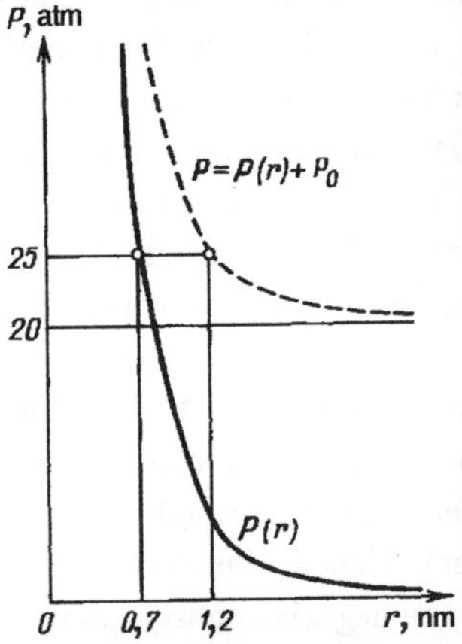

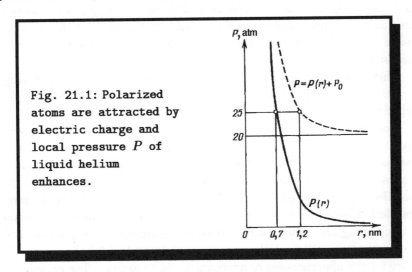

Fig. 21.1: Polarized atoms are attracted by electric charge and local pressure P of liquid helium enhances.

Well, as we already know, at low temperatures liquid helium solidifies at $25\,atm$. Since, as soon as the pressure in the vicinity of the positive helium ion reaches that critical value, a certain volume around it turns solid[a]. According to Fig. 21.1 for low external pressures this solidification takes place within approximately $r_0 = 0.7\,nm$ $(0.7 \cdot 10^{-9}\,m)$ from the ion. Therefore it gets "frozen" in a "snowball" formed by solid helium. Now, if electric field is applied, the "snowball" will start moving. But it's not going to move alone, sure enough, it will be accompanied by its new retinue, pulling along a whole "tail" of that extra density.

Consequently, the total mass of the positive charge carrier will include the three major contributions. The first one is the mass of the "snowball" itself, which is the product of the density of solid helium times the "snowball" volume at normal ambient pressure. This gives $32\,m_0$ ($m_0 = 6.7 \cdot 10^{-27}\,kg$, is the mass of helium atom). The mass of the following "retinue" turns out to be just slightly less — the mass of the tail of extra density pulled by the ion amounts to $28\,m_0$.

Besides these two, there must be one more mass added: when an object moves in liquid, there always occurs some displacement of masses of the liquid around it. This, of course, takes some energy. That is, to accelerate

[a]That, as you remember from Chapter 20, is rather flabby.

a body in liquid requires greater force than the same acceleration in vacuum. So in liquid an object behaves as if its mass was somewhat greater. This additional mass due to the motion of liquid layers is called the ***associated mass***[b]. For the "snowball" moving in the liquid helium at normal atmospheric pressure the associated mass turns to be $15m_0$.

Finally, after summing up, the total mass of the positive charge moving in the liquid helium equals $75m_0$, the value which closely agrees with the experimentally measured one.

You see that the concepts of classical physics may deal successfully with the theory of positive charge carriers in liquid helium. But it is not so easy with negative ones. First of all, it happens that there are no negative ions in liquid helium at all (although a few of negatively charged molecular ions He_2^- may be formed, yet they don't play any noticeable part in charge transfer). Hence, the electron is still the only remaining contestant for the role of the negative charge carrier. However, it catastrophically misses most of the mass required from the experimental data. And here is exactly the place for the idiosyncrasies of quantum world to appear. The experiment shows, that electron, whom we have been so stubbornly intending for the negative charge carrier can't even freely penetrate into liquid helium.

To make sense of all this, we will have to digress and touch a little on the structure of atoms with several electrons. There is a paramount principle, unquestionably reigning in the microworld, determining behavior of groups of identical particles. When applied to electrons, it is called the ***Pauli***[c] ***exclusion principle.*** According to this rule, no two electrons can occupy the same quantum state at the same time. And we shall show that this explains the observed "aversion" of helium atoms towards free electrons and the troubles the latter run in when they try entering the liquid helium.

The energy of electron in atom, as we've noted already, can have only certain quantum values. And what's important is that for each such value of energy there are several corresponding states available for electrons, varying by the character of their motion in atom (for instance, by the electron orbit shape or, in quantum tongue, the shape of probability cloud which defines the spread of electron in space, see Fig. 20.3). States of the same

[b]We have encountered this concept already in Chapter 12 (see page 103).

[c]Wolfgang Pauli, (1900–1958), Austrian physicist in the US; works in quantum mechanics, quantum field theory, relativity and other fields of theoretical physics; Nobel Prize 1945.

energy compose a so-called ***shell***. According to the Pauli principle, when the number of electrons in atom grows (as the atomic number goes up), they do not get "crammed" in the same states but fill one by one new available shells.

The first shell, corresponding to the lowest possible energy, must be occupied first. Located real close to the atomic nucleus, it can take in only two electrons. Thus, in helium, which is the second in the periodic table, that first shell is completely filled. There is no choice for the third electron other than to stay sufficiently far from the nucleus. When such an "undue" electron approaches a helium atom within a distance of the order of its radius, there appear repulsion forces precluding the further nearing.

Therefore, some "entrance work" is needed to ram an errant electron into the bulk of helium. Three Italian physicists, Carreri, Fasoli and Gaeta, proposed an idea that as the electron entering helium can't get too close to atoms, it shoves them away and thereby forms a spherically symmetrical cavity, some sort of a "bubble" Fig. 21.2. And this bubble with the electron scurrying inside is indeed the negative charge carrier in liquid helium.

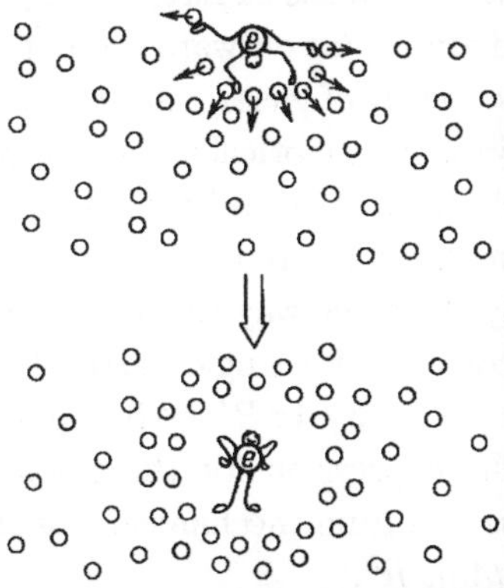

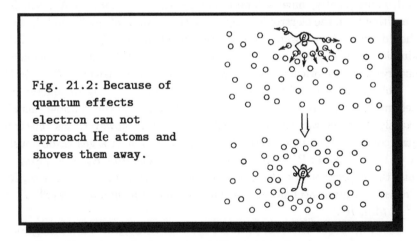

Fig. 21.2: Because of quantum effects electron can not approach He atoms and shoves them away.

The size of the bubble could be rather easily estimated. The repulsion between electrons and helium atoms, should decrease with distance. On the other hand, at large distances electrons should act on helium atoms in exactly the same way the positive ions do, that is to polarize them. So, far enough the interaction of electrons with helium atoms should be the same attraction as for the considered above situation with "snowballs".

Hence, when approaching a "bubbles" with the electron captured inside, the incremental pressure in helium rises following the law portrayed in Fig. 21.1.

Yet, under normal conditions the pressure at the boundary of the "bubble" still remains far less than 25 atm because of the comparatively large size of the "bubble". Besides this pressure, resulting from the increased of density of the polarized helium, there is a force of surface tension. It acts at the bubble's border too and is directed in the same way, to the center of the "bubble". What would then balance these two external forces not allowing our "bubbles" to collapse? It turns out to be that the needed counteraction is created by that very same "imprisoned" electron.

Indeed, according to the uncertainty principle, which we have discussed in Chapter 20, the accuracy of measuring momentum of electron is directly related to the uncertainty of the electron's position in space, being $\Delta p \sim \hbar / \Delta x$. In our case the uncertainty in location of the electron is naturally defined by the size of the "bubble", that is $\Delta x \sim 2R$. So, the rushing inside captive electron should posses a momentum of the order of $\hbar / 2R$ and, consequently, have a kinetic energy $E_k = p^2/2m_e \sim \hbar^2 / 8m_e R^2$. Resulting from the electron's collisions with the walls of the "bubble", there should arise some outward pressure (remember the principal equation of the kinetic theory of gases, relating the pressure of gas P to the average kinetic energy of its chaotically moving particles and their density: $P = \frac{2}{3}nE_k$). This pressure could very well balance the forces trying to squeeze our helium "bubble". In other words, electron confined in the "bubble" acts in exactly the same way as gas isolated in a reservoir, however, this "electron gas" consists of a single particle! The density of such a gas is, obviously, $n = 1/V = 3/4\pi R^3$. After plugging that value and $E_k \approx \hbar^2 / 8m_e R^2$ into the expression for the pressure, we find that $P_e \approx \hbar^2 / 16\pi m_e R^5$. Precise quantum mechanical calculations lead to a similar answer[d]: $P_e \approx \pi^2 \hbar^2 / 4m_e R^5$.

As long as the external pressure remains small, the dominating force trying to squeeze the "bubble" will be the surface tension, $P_L = 2\sigma / R$, see Chapter 10. Hence, equating $P_e = P_L$, one can easily estimate the

[d]The origin ot the discrepancy is that the electron prefers to stay in the middle of the cavity rather than near the repelling walls. This effectively reduces the uncertainty of coordinate and enhances the pressure.

radius of a stable "electron bubble" in liquid helium:

$$R_0 = \left(\frac{\pi^2 \hbar^2}{8 m_e \sigma} \right)^{\frac{1}{4}} \approx 2\,nm = 2 \cdot 10^{-9}\,m.$$

We see now that the negative electrical charge in liquid helium is carried by "bubbles" with electrons "ensnared" inside.

The total mass of such carriers can be calculated in the same manner as was done for the "snowballs". Yet, now the "bubble" itself weighs almost nothing for the mass of the electron inside is negligibly small compared to the mass of the liquid dragged by the "bubble" (the "retinue") plus the associated mass. So, the net mass of the carrier would be equal to the sum of the associated mass and the mass of the "tail" pulled by the drifting "bubble". Because of the rather large size of the "bubble", its resulting mass, $245\,m_0$, turns out to be much greater than that of the "snowballs".

Now, let us consider how an increase in external pressure P_0, will effect the properties of the charge carriers. Fig. 21.1 depicts the dependence of the total pressure (including the external one), $P = P(r) + P_0$, in the vicinity of an ion in liquid helium versus distance from the ion for $P_0 = 20\,atm$. Such dependence for an arbitrary value of $P_0 < 25\,atm$ can be graphed by simply shifting the curve for $P_0 = 0$, along the P-axis. As the picture shows, the higher the external pressure is, the further from the ion the total pressure attains $25\,atm$. Hence, with the growth of the external pressure, the "snowball" behaves as if it was rolling down a snowy slope: it promptly lumps up on itself the "snow" — solid helium, turning bigger and bigger. The relation between the size of the snowballs and the value of external pressure $r(P_0)$, is presented in Fig. 21.3.

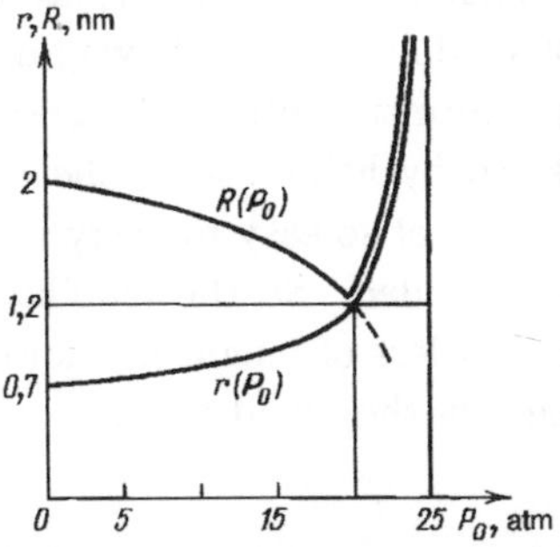

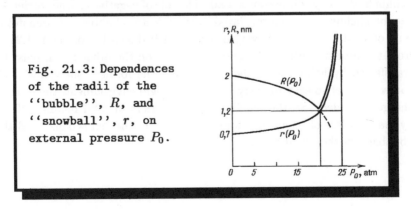

Fig. 21.3: Dependences of the radii of the "bubble", R, and "snowball", r, on external pressure P_0.

And what about the "bubble"? How does it behave as P_0 goes up? Well, for some while, like any bubble in liquid, it submissively shrinks with the surrounding pressure growing. Its radius $R(P_0)$ decreases as the upper curve in Fig. 21.3 shows. And yet, further, at $P_0 = 20\,atm$ the curves for $r(P_0)$ and $R(P_0)$ cross, meaning that at this point the sizes of the "bubble" and the "snowball" become the same and equal $1.2\,nm$. We know already the future fate of the "snowball": with rise of P_0 it will rapidly grow at the expense of the solidifying on its surface helium. But what the "bubble" is to do — continue to shrink following the dashed line in Fig. 21.3?

No way! At this very moment, our "bubble" finally shows its real character. As the applied pressure continues to go up, the bubble begins to act as a "snowball": it gets covered with icy crust of the solid helium. Indeed, according to Fig. 21.1 at $P_0 = 20\,atm$ the total pressure on the surface of the "bubble" becomes equal to $25\,atm$, reaching the solidification point for liquid helium. The internal radius of the bubble, protected now by its fancy "icy attire", ceases changing and remains approximately the same regardless of the further increase of pressure, whereas its external radius equals that of the "snowball" at the corresponding pressure.

Thus, at external pressures greater than $P_0 = 20\,atm$, our "bubbles" become coated by the ice shell and begin to somehow resemble nuts. With that difference, though, that the kernel in the "nut" is of a quite peculiar nature — it's an electron rushing chaotically inside the shell formed of solid helium.

One last thing worth mentioning here. As P_0 approaches $25\,atm$, the external radii of both the "nuts" and the "snowballs" continue to grow bigger and bigger (trying, in principle, to reach infinity). Finally all the helium in the reservoir becomes solid. The role of negative charge carriers in the solid helium is, therefore, played by the "electron bubbles", frozen in the bulk of solid helium and having inherited their dimension of about $1.2\,nm$ from the former "nuts". The positive charge, on the other hand, must be transferred by helium ions, the remnants of the former "snowballs". Of course, it's not so easy to carry anything in rigid environment, the electric charges included. So the mobility of the carriers in solid helium will be by many orders of magnitude lower than that of the "snowballs" and the "bubbles" in the liquid phase.

Chapter 22

The Superconductivity Passion at the end of the Millenium

Probably almost all our readers have heard of superconductivity. This phenomenon consists in the abrupt disappearance of electrical resistance of some pure metals and alloys at low temperatures. Almost all over the last century "low temperatures" meant the range of 10–20 K, that is 10–20 degrees above the absolute zero temperature ($-273.15^\circ C$). In order to cool to this low temperatures a sample is usually placed into liquid helium that at normal pressure boils at 4.2 K and, as you already know, does not freeze down to the absolute zero. Throughout the century physicists and chemists in many laboratories all over the world have been looking for compounds which become superconducting at high enough temperatures and could be cooled, for instance, by comparatively cheap and widely available liquid nitrogen. So you understand that the discovery of high-temperature superconductors, whose resistance becomes zero at temperatures above 100 K, was met as the greatest event in physics of recent years. Really, the practical significance of this discovery can be compared to that of magnetic induction at the beginning of the 19-th century. It ranks with the discovery of uranium fission, the invention of the laser, and the discovery of the unusual properties of semiconductors in the 20-th century.

22.1 Starting from the end

The beginning of this exciting new stage in the development of superconductivity was the work by K. A. Müller and T. J. Bednorz at IBM's lab in Switzerland. In the winter of 1985–86 they managed to synthesize a compound of barium, lanthanum, copper, and oxygen — the so-called metal oxide ceramic La – Ba – Cu – O, a compound which had superconducting properties at the record at that (still recent) time temperature of $35\,K$. The article, cautiously titled "The Possibility of High-Temperature Superconductivity in the La – Ba – Cu – O System", was turned down by the leading American journal ***Physical Review Letters.*** The scientific community had gotten tired over the past 20 years of receiving sensational reports about the discovery of high-temperature superconductors that turned out to be false, so it decided to save a trouble. Müller and Bednorz sent the article to the German journal ***Zeitschrift fur Physik.*** Now that the news about high-temperature superconductivity finally have been heard and research is being done in hundreds of laboratories, every article devoted to the new phenomenon would start from a reference to this article. But in the fall of 1986 it passed practically unnoticed. Just one Japanese group checked the result and verified it. Soon the phenomenon of high-temperature superconductivity was corroborated by physicists in the United States, China, and the Soviet Union.

At the beginning of 1987 the whole world was in a fever, searching for new superconductors and investigating the properties of those already discovered. The critical temperature T_c increased quickly: it was $T = 45\,K$ for La – Sr – Cu – O and it reached $52\,K$ for La – Ba – Cu – O under pressure. Finally in February 1987 the American physicist Paul Chu got the idea to imitate the effect of external pressure by substituting La atoms by the smaller atoms of Y that is the next in the Mendeleev table column. The critical temperature of the compound Y – Ba – Cu – O (see Fig. 22.1) broke the fabulous "nitrogen barrier", having reached $93°\,K$. This was a long awaited triumph but far not the end of the story. In 1988 a five-component compound of the type Ba – Ca – Sr – Cu – O with the critical temperature $110\,K$ was synthesized and a little later its mercury and tallium analogues with the critical temperature $125\,K$ appeared. The maximum critical temperature under the pressure of $30\,hPa$ of the mercury record-breaker looks impressive even on Celsius scale: it makes $-108°\,C$!

The discovery of high-temperature superconductivity is unique in modern physics. First, it was discovered by just two scientists with very modest tools. Second, the compounds include easily accessible elements. As a matter of fact these superconductors can be made in a high school chemistry lab in a day. What a contrast with discoveries in other areas of physics — for instance, high-energy particle physics. There the investigations are carried out by large teams of scientists (the list of authors takes a whole

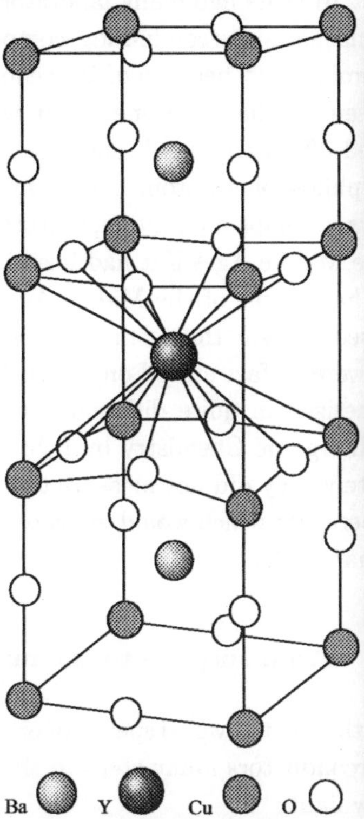

Fig. 22.1: Structure of superconducting $Y Ba_2 Cu_3 O_7$

page in a journal article), and the equipment costs millions of dollars. This discovery was a cause for optimism: the time of lone investigators in physics hasn't passed! And though the discovery had been anticipated for 75 years, it caught everyone by surprise. Theorists could just shrug their shoulders and shrugged them even harder as the critical temperature went up.

So was the discovery by Bednorz and Müller a fluke or the destiny? Could the discovered compound, with its unique properties, has been synthesized earlier? How difficult is to answer these questions! We have long been accustomed to the fact that everything new is obtained on the edge of the impossible by using unique equipment, superstrong fields, ultralow temperatures, superhigh energies... There is nothing of the kind here. It isn't too difficult to "bake" a high-temperature superconductor — a qualified alchemist of the Middle Ages could have managed it. It's worth recalling that about 10 years ago many laboratories of the world intensively investigated an unusual superconducting compound. This substance was called "the alchemic gold" because of its yellow luster and high density, which made it resemble the noble metal. It was synthesized by medieval alchemists, passed off as true gold, and advertised as the result of successfully using the "philosopher's stone". Alchemic gold is a complex compound, and who knows, perhaps a high-temperature superconductor could have been baked in the Middle Ages if it had been blessed with a golden luster.

The dreams of the Middle Ages may take us too far. But you will be amused to hear that some of present-day high temperature superconductors were in fact stored on lab shelves since 1979! At that time they were synthesized in quite different connection at Moscow Institute of General and Inorganic Chemistry by I. S. Shaplygin and his collaborators. Unfortunately they did not measure the conductivity of the compositions at low temperature which would indicate the new phenomenon. The discovery did not happen[a]...

22.2 From surprise to the understanding.

Now that all the world speaks of properties and prospects of high-temperature superconductors many steps in the history of superconductivity appear in a new light.

Superconductivity, one of the most interesting and unusual phenomena in solid-state physics, first became known on April 28, 1911, at a meeting of the Royal Academy of Sciences in Amsterdam, when the Dutch physicist Heike Kamerlingh Onnes[b] reported a recently discovered effect: the complete disappearance of electrical resistance of mercury cooled by liquid helium to $4.15\,K$. Though no one expected this discovery, and it contradicted the existing classical electron theory of metals, the fact that it was Kamerlingh Onnes who discovered superconductivity was not accidental. Actually, he was the first scientist who managed to solve the most complicated scientific and technical problem of the time: obtaining liquid helium (which boils at $4.16\,K$). This allowed scientists to peek into the unknown world of temperatures close to the absolute zero. Kamerlingh Onnes imme-

[a]A superconductor baked in the Middle Ages had been doomed to oblivion for the same reason: there was no liquid helium at hand! —*A. A.*

[b]H. Kamerlingh Onnes, (1853–1926), Dutch physicist; Nobel Prize 1913.

diately tried to apply the new experimental means and to investigate the low-temperature behavior of pure metals. This was the time of hot theoretical debate whether resistance of pure metals turns to zero or remains finite at absolute zero. Being the advocate of the first side Kamerlingh Onnes was clearly satisfied by the result that he had obtained for mercury. But soon he realized that the vanishing of resistance at finite temperature is an effect quite different from the expected one.

We'd like to emphasize that the resistance of a sample in the superconducting state is equal to zero not approximately but exactly. That's why electric current in a closed circuit can circulate as long as you like without damping. The maximal duration of a nondamped superconducting current recorded in England was about two years. (The current in the ring would have circulated up till now but for a strike of transport workers which caused a break in the supply of liquid helium to the laboratory.) Even after the two years, no damping of the current was detected.

Very soon superconductivity was discovered not only in mercury but in other metals as well. The prospects for practical applications of the discovered phenomenon seemed unlimited: power transmission lines without waste, superpowerful magnets, electric motors, new types of transformers... But there were two obstacles.

The first were the extremely low temperatures at which superconductivity was observed in all materials known by the time. To cool conductors to these temperatures, scarce helium is used (its stocks are limited, and even now producing a liter of liquid helium costs some dollars). This makes many projects to apply superconductivity simply unprofitable. The second obstacle discovered by Kamerlingh Onnes was that superconductivity had turned out to be rather sensitive to magnetic fields and to the value of current. In fact, it was destroyed by strong fields.

The next fundamental property of the superconducting state discovered in 1933 was the Meißner-Ochsenfeld effect: the complete expulsion of magnetic field from the volume of the superconductor. But again experimental investigations were complicated by the need to work with scarce liquid helium — before the World War II it was produced in about 10 laboratories throughout the world (the two of those were in the Soviet Union).

The fundamentals of superconductivity stayed absolutely out of reach of classical theory of metals whereas the quantum one was in embryo. The so-called two-liquid model suggested a coexistence of two types of electrons in superconducting metals: normal electrons interact with lattice but superconducting ones for some reason don't. This assumption let brothers London[c] to write down the equations of electrodynamics of superconductors that described the Meißner effect and some other features. Still the microscopic mechanism of superconductivity remained a mystery.

In 1938 P. L. Kapitza[d] discovered ***superfluidity***. It turned out that at temperatures below 2.18 K liquid helium can flow through whatever thin capillary tubes without any viscosity. The theoretical explanation of this phenomenon by L. D. Landau[e] gave rise to hopes that the theory of superconductivity was in the offing. It turns out that helium atoms at low temperature behave like quantum particles with whole spin and get accumulated at the lowest energy level (the ***Bose-condensation***). Landau has shown that a gap that appears as the result in the spectrum of excitations makes possible the superfluid state. Discussing this macroscopic revelation of the entirely quantum effect Landau called helium "a window to the quantum world".

A straightforward extension of these ideas to superconductivity failed. The reason was that electrons are particles with spin one-half (so-called *fermions*) and behave absolutely unlike helium atoms which possess a whole spin being ***bosons***. In quantum system of electrons excitations with zero energy may appear even at zero temperature and the Landau criterion of superfluidity does not hold.

The natural desire to reduce the problem to that already solved inspired the idea to prepare of two fermions a composite boson with a whole total spin and after that to effect the Landau superfluidity scenario. However this was opposed by Coulomb[f] repulsion of electrons that was too strong in spite of screening that occurs in electroneutral metal.

Ten years later, in 1950, the discovery of the ***"isotopic effect"*** first indicated the connection between superconductivity and the crystal lattice of the metal. Measurements of the critical temperature of lead proved that it depended on the mass number of the isotope under testing. Thus superconductivity ceased being a purely electronic phenomenon. A little later

[c]H. London, (1907–1970), British physicist; F. London, (1900–1954), American physicist; specialists in low temperature physics.

[d]P. L. Kapitza, (1894-1984), Russian physicist; Nobel Prize 1978.

[e]L. D. Landau, (1908–1968), Russian physicist; Nobel Prize 1962.

[f]C. A. de Coulomb, (1736–1836), French physicist and inventor.

Frölich[g] and Bardeen have independently demonstrated that interaction of electrons with lattice oscillations ***(phonons)*** may lead to attraction. This could in principle overcome the electrostatic repulsion but one had to keep in mind the huge kinetic energies of electrons. At the first sight those should break the just found weak coupling. Composite bosons did not work out.

In the same 1950 with the help of experimental data and theoretical achievements of solid-state physics, based on quantum mechanics and statistical physics, Ginzburg[h] and Landau (USSR) developed a phenomenological theory of superconductivity, known as the Ginzburg-Landau theory. It proved so successful and predictive that even now it remains a powerful research tool despite that the 50 elapsed years were marked by the creation of the microscopic theory of superconductivity.

In 1957 the American scientists John Bardeen, Leon Cooper and Robert Schrieffer[i] put together the mentioned above ideas and hints and created a consistent microscopic theory of superconductivity. It was found that superconductivity is indeed linked with the appearance of a peculiar attraction of electrons in metals. This is an utterly quantum phenomenon.

We have already mentioned that ground state of fermionic system is characterized by big kinetic energies of electrons. Luckily those do not prevent binding of low-energy excitations of the system that behave like ***quasiparticles***. They have the same electric charge e as electron and some effective mass but their energy may be whatever small. The attraction brings on a rearrangement of quasiparticle spectrum and the long-awaited gap that was so crucial for the Landau superfluidity criterion opens at last.

The origin of the attraction may be understood with the help of a far analogy with two balls lying on a rubber rug. If the balls are far from each other, each of them deforms the rug, making a little depression. But if we put a ball on the rug and place another one near the first, their holes will join, the balls will roll down to the bottom of the combined valley and lie together. In metals the mechanism is realized by deformations of crystal lattice. At low temperatures some quasiparticles (usually they are called, just the same, electrons) form a sort of bound pairs. These are

[g]H. Frölich, (born 1905), British physicist.

[h]V. L. Ginzburg, (born 1916), Russian physicist and astrophysicist.

[i]J. Bardeen, (born 1908), American physicist; Nobel Prize 1956, 1972 (!).
L. Cooper, (born 1930), and R. Schrieffer, (born 1931); American physicists; Nobel Prize 1972.

called "Cooper pairs" after the man who discovered the binding. The size of the pairs on the atomic scale is really quite large, reaching hundreds and thousands of interatomic distances. According to the graphic comparison suggested by Schrieffer, they should be envisaged not as a double star composed of electrons but rather like a couple of friends in a discotheque who either come together or dance in different corners of the hall, separated by dozens of other dancers.

You see that it took almost half a century since the discovery to gain cardinal progress in understanding the nature of superconductivity and to develop the consistent theory. This period may be considered to be the first stage of superconductivity studies.

22.3 Chasing high critical parameters.

The creation of the theory of superconductivity was a powerful impulse to investigate it in earnest. Without fear of overstatement, we can say that great progress has been achieved in producing new superconducting materials in the subsequent years. The Soviet scientist A. A. Abrikosov's discovery of an unusual superconducting state in a magnetic field played a significant role in this development. Before then magnetic field was thought to be incapable of penetrating the superconducting phase without destroying it (which is actually true for most pure metals)[j]. Abrikosov theoretically proved that there was another possibility: under certain conditions magnetic field could penetrate into superconductor in the form of current vortices. The core of the vortex turned into the normal state but the periphery remained superconducting! Depending on the behavior in magnetic field, superconductors were divided into two groups: superconductors of the first type (old) and those of the second type (discovered by Abrikosov). It's important that superconductor of the first type can be changed into one of the second type if we "spoil" it by adding impurities or other defects.

A real hunt for superconducting materials with high critical fields and temperatures started. The ingenuity of the pursuers was really boundless. Arc welding, instant cooling and sputtering onto hot substrate were

[j]Strictly speaking this is true only for cylindrical specimens placed in a magnetic field parallel to the axis of the cylinder. If either the specimen is not a cylinder or a strong enough field is oriented differently the so-called intermediate state may be realized. It is formed by alternating macroscopic layers of superconducting and normal phases.

employed. The efforts resulted into the discoveries of, for example, the alloys $Nb_3\,Se$ and $Nb_3\,Al$ which have the critical temperature (that is the temperature of getting superconducting) $T_c = 18\,K$ and the upper critical field more than $20\ T$. The further progress was achieved lately with ternary compounds. Before the discovery of high-temperature superconductors the record value of the upper critical field ($60\ T$) belonged to the alloy $Pb\,Mo_6\,S_8$ with $T_c = 15\,K$.

Among superconductors of the second type, scientists managed to find compounds capable of carrying a high-density current and bearing gigantic magnetic fields. And although many problems had had to be solved before they could find practical application (the compounds were brittle, high currents were unstable), the fact remained: one of the two major obstacles to the widespread use of superconductors in technology was overcome.

But increasing the critical temperature still was problematic. If critical magnetic fields were increased thousands of times in comparison with Kamerlingh Onnes's first experiments, the changes in critical temperature weren't too encouraging: it only managed to reach $20\,K$. So for the normal operation of superconducting instruments the expensive liquid helium was still necessary. This was particularly vexing because a fundamentally new quantum effect, the "Josephson effect", had been discovered. This made it possible to use superconductors widely in microelectronics, medicine, instrumentation, and computers.

The problem of increasing the critical temperature was extremely acute. Theoretical evaluations of its peak value showed that in the context of normal phonon superconductivity (that is, superconductivity caused by electron attraction due to the interaction with crystal lattice), this temperature could not exceed $40\,K$. But the discovery of a superconductor with such a critical temperature would be a great success, since that could be achieved with relatively cheap and available liquid hydrogen (which boils at $20\,K$). It would open the era of "mid-temperature superconductivity". This stimulated attempts to modify existing superconductors and create new ones by traditional methods of material science. But the ultimate dream was to create a superconductor with a critical temperature of $100\,K$ (or, even better, above room temperature), which could be cooled by cheap and widely used liquid nitrogen.

The best result of the search was the alloy $Nb_3\,Ge$ with the critical temperature of $23.2\,K$. This record temperature was achieved in 1973 and stood for 13 years. Until 1986 the critical temperature couldn't be raised by even one degree. It seemed that the possibilities of the phonon mechanism of superconductivity had been exhausted. In view of this, in 1964 the american physicist Little and the Soviet scientist V. L. Ginzburg proposed the following idea: if the possibility of increasing the critical temperature is limited by the nature of the phonon mechanism of superconductivity then it should be replaced by some other one — that is, electrons should form Cooper pairs by means of some other, nonphonon, attraction.

During the last 20 years many theories were proposed, tens or hundreds of thousands of new substances were investigated in detail. In his work Little has drawn attention to quasi-one-dimensional compounds — long molecular conducting chains with side branches. According to theoretical evaluations, a noticeable increase in critical temperature could be expected there. Despite attempts of many laboratories throughout the world, such superconductors were not synthesized. But on the way physicists and chemists have made many wonderful discoveries: they obtained organic metals, and in 1980 crystals of organic superconductors were synthesized (the current record for the critical temperature of an organic superconductor is over $10\,K$). They managed to obtain two-dimensional layered metal-semiconductor "sandwiches" and at last magnetic superconductors where the former enemies, superconductivity and magnetism, coexisted peacefully. But there were no new prospects for high-temperature superconductivity.

By this time superconductors had extended the range of application, but the need to cool them with liquid helium remained the weak spot.

In the mid-1970-s strange ceramic compounds of the type Pb – Ba – O appeared as candidates for high-temperature superconductivity. In their electrical properties they were "poor metals" at room temperature but became superconducting not too far from absolute zero. "Not too far" means about 10 degrees below the record value of the time. But the new compound could hardly be called a metal. According to theory, the obtained value of critical temperature was not by any means low but surprisingly high for such substances.

This attracted attention to ceramics as "would be" high-temperature superconductors. Since 1983 Müller and Bednorz worked like alchemists with hundreds of different oxides, varying their composition, quantity, and conditions of synthesis. According to professor Müller they were led by some physical ideas that are now getting validated by the most complicated experimental studies of the new materials. In this painstaking way they stealthily approached the compound of barium, lanthanum, copper, and oxygen that showed superconducting properties at $35\,K$.

22.4 Quasi-two-dimensional superconductivity between antiferromagnetic and metallic states

A fair number of various chemical compounds exhibiting temperatures of the superconducting transition higher than the record of 1973 were synthesized by now. Chemical formulae of some of them that have critical temperatures above the liquid nitrogen boiling point are summarized in Figure 22.2.

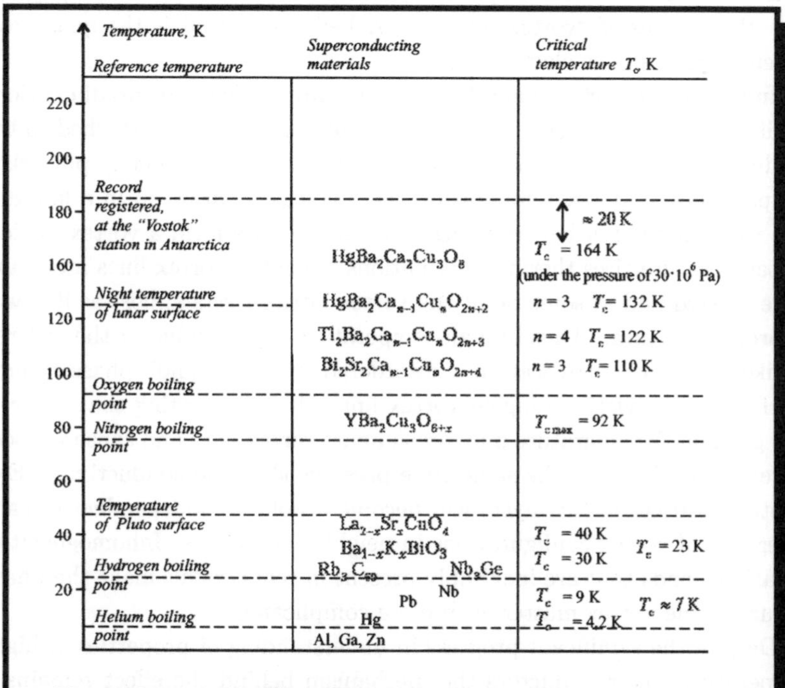

Fig. 22.2: HIGH T_c SUPERCONDUCTORS.

The feature shared by high-temperature superconductors is the layered structure. The best studied high-temperature superconductor is by now the compound $Y\,Ba_2\,Cu_3\,O_7$. Its crystal structure is illustrated in Fig. 22.1

on page 197. It is easily seen that atoms of copper and oxigen are arranged in planes interspaced by others atoms. As a result the conducting layers are separated by insulating ones and the motion of charge carriers (those are usually not electrons but holes) is ***quasi-two-dimensional.*** Namely, holes migrate freely within $Cu\,O_2$ layers although hops between the layers are comparatively rare. Cooper pairs are also localized in the planes.

Apparently the quasi-two-dimensional nature of the electron spectrum in high-temperature superconductors is a key to the understanding of the microscopic mechanism of this wonderful phenomenon. This has to be done yet. Nevertheless a brilliant phenomenological theory of the vortex state in high temperature superconductors is already at hand. It proved so rich with diverse effects that now, in fact, it constitutes a new realm of physics, *i. e.* the physics of "vortex matter". A bedrock of that is the quasi-two-dimensionality of the electron liquid.

Indeed, once electrons and Cooper pairs are confined in two dimensions Abrikosov vortices consist of elementary vortices which are attached to the conducting planes. These elementary vortices are known among physicists as "pancakes". At low temperature the "pancakes" draw up in a line due to a weak attraction between them. Then the lines form a vortex lattice. As temperature rises thermal fluctuations make the vortex lines more and more twisted and at some point the vortex lattice melts almost as if it was an ordinary crystal. Thus in high-temperature superconductor the ordered Abrikosov lattice gives way to a disordered "vortex liquid" phase formed by chaotically twisting tangled vortex lines. It is interesting that the further growth of temperature may break apart vortex lines and cause vortices to "evaporate" but at the same time preserve the superconductivity. Elementary vortices in the layers will become absolutely independent of each other and of vortex configurations in neighboring planes. Inhomogeneities of various kinds that are inevitably present in real crystals make the phase picture of the vortex matter even more complicated.

Despite the significant progress in understanding of properties of high-temperature superconductors the mechanism behind the effect remains a secret. None less than twenty conflicting theories proclaim having explained the high-temperature superconductivity but what we need is the only one that is true.

Some physicists believe that Cooper pairs in these superconductors are formed because of a magnetic fluctuation interaction of some sort. An indication may be the fact that the critical temperature and the concentration of free electrons drop down in crystals $Y\,Ba_2\,Cu_3\,O_{6+x}$ impoverished in oxigen, that is at $x < 1$, Fig. 22.3, (the right curve). At $x < 0.4$ one deals already with a dielectric but at temperatures low enough a magnetic ordering of copper atoms takes place. Magnetic moments of neighboring atoms become antiparallel and the total magnetization of the crystal stays zero. This type of ordering is well known in physics of magnetism where it is called ***antiferromagnetic ordering*** (see the left curve in Fig. 22.3; here T_N is the so-called Néel[k] temperature, that is the temperature of the transition to the antiferromagnetic state). One could believe that copper atoms retain the fluctuating magnetic moment in superconducting phase and in the long run that gives origin to the superconducting attraction of electrons. This mechanism leans on specific properties of copper atoms which, depending on the valence, are either magnetic or not. Presence of $Cu - O$ layers in all high temperature superconductors could be considered as an argument in favor of the theory. However quite recently the superconductivity of $W_3\,O\,Na_{0.05}$ at $90\,K$ was reported. Exact composition of the superconducting phase is not known yet but for sure there are no "magic" copper atoms in it. Moreover, none of the elements in the formula of the new high temperature superconductor shows magnetic properties.

In other theories physicists try to generalize in one or another way the classical theory of superconductivity, revise the very basics of theory of metallic state, "crossbreed" superconductivity and ferromagnetism in spaces of higher dimension, separate spin and charge of carriers, concoct Cooper pairs in advance at temperatures higher than critical and undertake other attempts to explain unusual properties of the high-temperature superconductors in a universal manner.

The challenge of nature waits for the answer, theoretical community can not come to accord. On the one hand, this may confirm the famous comparison with a chorus of deaf where everyone performs his own part and does not care of the others. But on the other, it may happen that time has not come and the correct theory is not formulated yet.

[k]L. Néel, (born 1904), French physicist; Nobel Prize 1970.

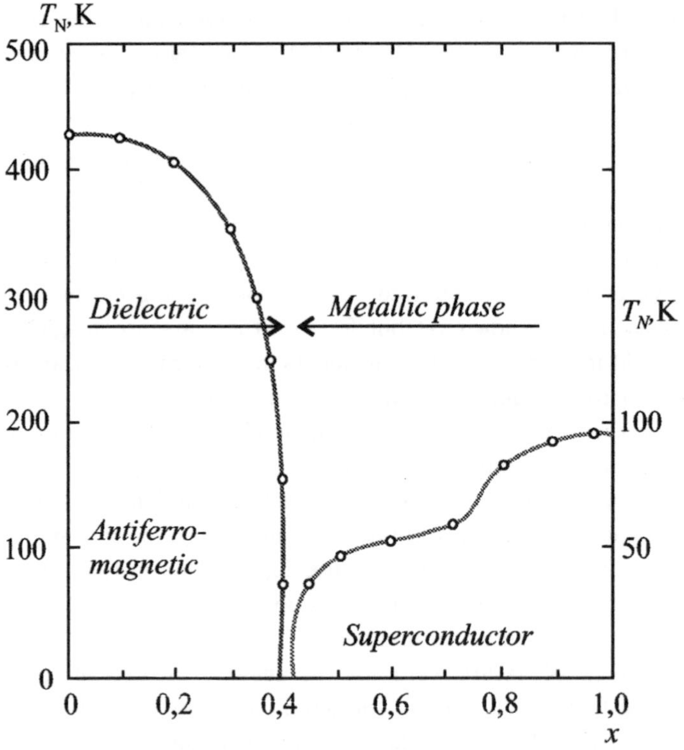

Fig. 22.3: Phase diagram of $YBa_2Cu_3O_{6+x}$

Chapter 23
What is SQUID?

23.1 The quantization of magnetic flux

In microworld, that is the world of atoms, molecules and elementary particles, many physical quantities may take only definite discrete values. Physicists say that they are quantized. (We have already mentioned that according to the Bohr rule energies of electrons in atoms are quantized[a] Macroscopic bodies consist of big collectives of particles and chaotic thermal motion leads to averaging of physical quantities. This smears little steps and conceals quantum effects at macroscopic level.

Now, what if the body is cooled to a very low temperature? Then arrays of microparticles can move in accord and reveal quantization at macroscopic scales. A bright example is the fascinating phenomenon of quantization of magnetic flux.

Everyone who has studied laws of electromagnetic induction knows what is the magnetic flux through a closed contour:

$$\Phi = B\,S,$$

where B is the value of magnetic induction and S is the area encircled by the contour (for simplicity let the field be normal to its plane). Nevertheless it will be a discovery for many that magnetic flux produced by superconducting current, say, in a ring may assume only discrete values. Let us try to understand this at least superficially. Presently it suffice to believe that microparticles are moving along quantum orbits. This simplified image

[a]See page 178.

often substitutes probability clouds in blackboard discussions.

The motion of superconducting electrons in the ring, Fig. 23.1, resembles that of electrons in atoms: it seems that electrons follow gigantic orbits of the radius R without any collisions. Therefore a natural assumption is that the motion obeys the same rule as in atoms. The Bohr postulate states that only certain orbits of electrons are stationary and stable. They are selected by the following quantization rule: products of the momentum of an electron mv and the radius of the orbit R (this quantity is called the angular momentum of the electron) form a discrete sequence:

$$mv\,R = n\,\hbar. \tag{23.1}$$

Here n is a natural number and $\hbar$ is the minimal increment *(quantum)* of the angular momentum is equal to the Planck constant $\hbar$. We have already met it when talking about the uncertainty relation. It turns out that quantization of all physical quantities is determined by this universal constant.

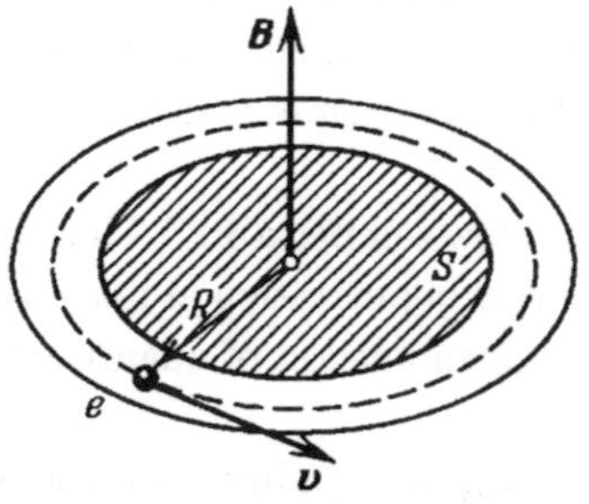

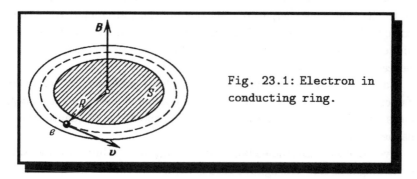

Fig. 23.1: Electron in conducting ring.

Let us find the value of the magnetic flux quantum. Consider a single electron and let the magnetic flux through the ring gradually increase. As you know the electromotive force of induction appears:

$$\mathcal{E}_i = -\frac{\Delta\Phi}{\Delta t},$$

and the strength of the electric field is

$$E = \frac{\mathcal{E}_i}{2\pi R} = -\frac{\Delta\Phi}{2\pi R\,\Delta t}.$$

By the second Newton law the acceleration of a charged particle is:

$$ma = m\frac{\Delta v}{\Delta t} = -\frac{e\,\Delta\Phi}{2\pi\,R\,\Delta t},$$

where e is the electric charge. After the obvious cancellation of Δt we obtain:

$$\Delta\Phi = -\frac{2\pi\,m\,\Delta v\,R}{e} = -\frac{2\pi}{e}\Delta\,(m\,v\,R).$$

You see that magnetic flux across the ring is proportional to the angular momentum of electrons[b]. According to the Bohr quantization rule (23.1) angular momentum may take only discreet values. This means that magnetic flux through a ring with superconducting current must be quantized as well:

$$m\,v\,R = n\,\hbar \qquad \text{and} \qquad \Phi = (-)\frac{2\pi}{e}\,n\,\hbar. \tag{23.2}$$

The value of the quantum is extremely small ($\sim 10^{-15}\,Wb$) but 20th-century technique makes possible to observe magnetic flux quantization. The studies were carried out in 1961 by the Americans Deaver and Fairbank. The only difference was that the superconducting current circulated not in a ring but in a hollow superconducting cylinder. The experiment confirmed that the magnetic flux through the cylinder changed stepwise but the measured value of the quantum was twice less than that obtained above. The modern theory of superconductivity gives the answer. Remember that in superconducting state electrons join into the Cooper pairs with the charge $2e$. Superconducting current is a motion of these pairs. Therefore the correct value of the magnetic flux quantum Φ_0 is obtained when substituting into the formula (23.2) the electric charge $2e$ of a pair:

$$\Phi_0 = \frac{2\pi\,\hbar}{2e} = 2.07 \times 10^{-15}\ Wb.$$

This is the way to recover the factor two. We were not the first to miss it. The english theoretician F. London had lost it as well. He predicted the magnetic flux quantization already in 1950, long before the nature of superconducting state was understood.

[b]The relation between encircled magnetic flux and angular momentum of electrons is valid both in classical and quantum physics. —*A. A.*

It is worth saying that our derivation of the magnetic flux quantization certainly is too naive. It is rather surprising that we contrived to obtain the right meaning of the quantum this way[c]. In fact superconductivity is a complicated quantum effect. Those who want really comprehend it have a long and hard way ahead of them. It demands many years of resolute but rewarding work.

23.2 Josephson effect

Let us turn to another quantum superconducting phenomenon that lay a cornerstone for several unrivaled measuring methods. The Josephson effect was discovered in 1962 by a 22-year-old British graduate student and brought him the Nobel Prize for the theoretical prediction 11 years later[d].

Imagine a glass plate (that is called a substrate) supporting a superconducting film. (Usually the superconducting material is sputtered in vacuum.) The surface of the film has been oxidized and the oxide forms a thin dielectric layer on it. Finally the superconductor was sputtered once again. The final outcome is a so-called superconducting sandwich interlaid by a thin insulating sheet. Sandwiches are widely used in observations of the Josephson effect. For convenience the two thin superconducting strips usually cross each other, see Fig. 23.2.

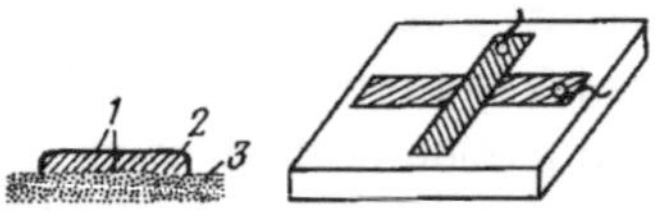

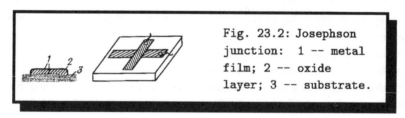

Fig. 23.2: Josephson junction: 1 -- metal film; 2 -- oxide layer; 3 -- substrate.

We shall begin with the case when the metallic layers are in normal, nonsuperconducting state. Is it possible for electrons to pass from one metallic film into another, Fig. 23.3, *a*?

[c]The weak points of our derivation are: first, it is impossible to change the magnetic flux in a superconducting ring due to the "freezing" (see later); second, superconducting pairs form a quantum collective state and there is no way to pick out a single pair. — *A. A.*

[d]B. D. Josephson, (born 1940), British physicist; Nobel Prize 1973.

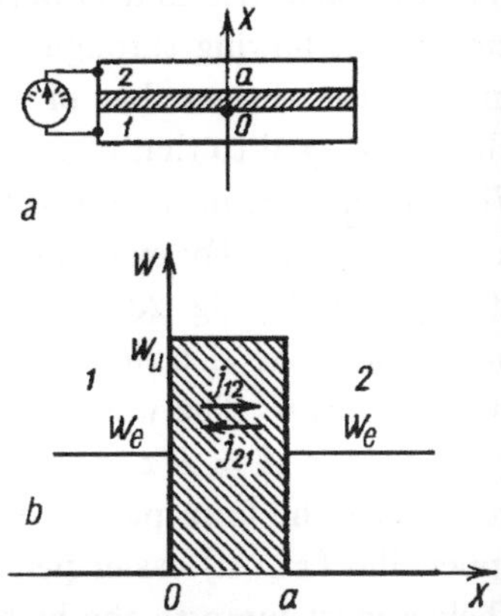

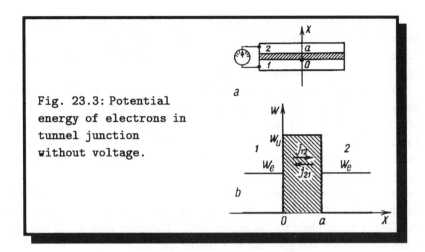

Fig. 23.3: Potential energy of electrons in tunnel junction without voltage.

From the first sight, not, because of the dielectric in between. The dependence of the electron energy versus the x-coordinate (X-axis is perpendicular to the plane of the sandwich) is plotted in Fig. 23.3, *b*. Electrons in metal move freely and their potential energy is zero. Potential energy of electrons in dielectric, W_u, surpasses their kinetic (and total) energy in metal W_e. The work to be done by electrons when exiting to the dielectric[e] is $W_u - W_e > 0$. Therefore one says that electrons in the two films are separated by the potential barrier of the height $W_u - W_e$.

In case that electrons obeyed the laws of classical mechanics the barrier would be insuperable. But electrons are microparticles and specific laws of microworld permit many things that would be ruled out for bigger bodies. For example neither man nor electron are able to mount a barrier higher than their energy. But electron may simply penetrate through it! As if it tunneled under a mountain when the energy was not enough to climb it. This is called the tunnel effect. Of course you should not take this literally like really digging a hole. The true explanation comes from wave properties of microparticles and their "spreading" in space. Real deep understanding of that requires good command of quantum mechanics. But the truth is that with some probability electrons can pass through the dielectric from one metal film to another. The probability increases for smaller heights $W_u - W_e$ and widths a of the barrier.

[e]This resembles the heat of evaporation that is the work done when extracting a molecule from liquid.

Once the dielectric film is permeable for electrons we may ponder of electric current flowing through it. At the moment this so-called tunnel current is zero: the number of electrons coming to the upper electrode from below is equal to that of going back.

What should we do to make the tunnel current nonzero? Simply to break the symmetry. For example let us connect the metal films to a battery with the voltage U, Fig. 23.4, *a*. Then the films will act like two plates of a capacitor and the electric field of the strength $E = U/a$ will set up in the dielectric layer. The work done when moving a charge e a distance x along the field is $A = F\,x = eE\,x = eU\,x/a$ and the potential energy of electrons takes the form plotted in Fig. 23.4, *b*. Evidently electrons from the upper film ($x > a$) easier penetrate the barrier because those moving from below must jump to the higher level. Therefore even small voltages break the balance and give rise to a tunnel current.

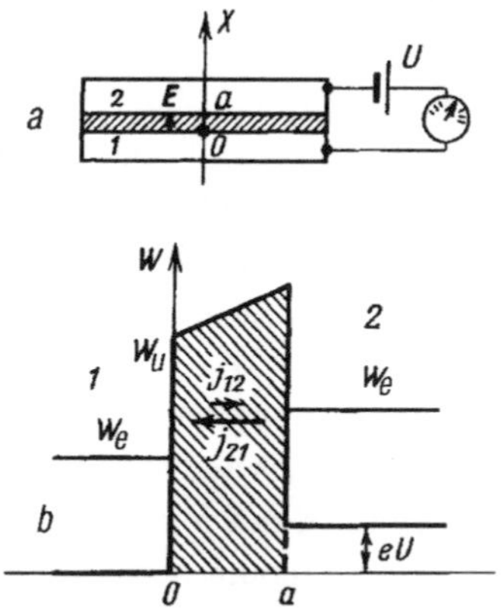

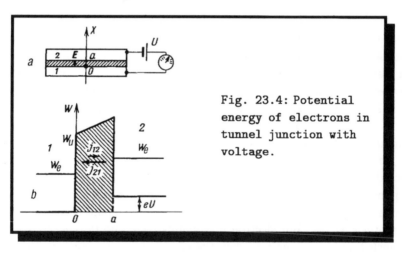

Fig. 23.4: Potential energy of electrons in tunnel junction with voltage.

Tunnel junctions of normal metals are used in electronic devices but don't forget that our aim have been practical applications of superconductivity. The next step is to assume that the metal strips separated by the insulating layer are superconducting. How behaves the superconducting tunnel junction? It turns out that superconductivity leads to quite unexpected results.

We said that electrons in the upper film possess the surplus energy eU with respect to the lower one. Upon coming down they must dump the energy and come to equilibrium with others. This was not a problem in normal state: several collisions with crystal lattice would redistribute the extra energy and convert it to heat. But if the film is superconducting this way is not acceptable! It remains to emit the energy in the form of quantum of electromagnetic radiation. The energy of the quantum is proportional to the applied voltage U:

$$\hbar\omega = 2eU.$$

You see that the electric charge in the right hand side is twice that of electron. This indicates that tunnelling of superconducting pairs takes place.

This was the dazzling prediction by Josephson. Applying constant voltage to superconducting tunnel junction (sometimes called the Josephson junction) brings about generation of electromagnetic radiation. The first experimental observation of this effect was performed in 1965 by I. M. Dmitrienko, V. M. Svistunov and I. K. Yansons in Kharkov Physical-Technical Institute of Low Temperatures.

The first that comes to mind is to use the Josephson effect for generation of electromagnetic waves. However it is rather difficult to extract the radiation from the narrow space between the superconducting films (this was a serious obstacle to experimental observation of the effect). Besides the emission is too weak. Now Josephson elements are used mainly as detectors of electromagnetic radiation being the most sensitive ones in certain frequency ranges.

This application exploits the resonance between the frequency of the external (registered) wave and the proper frequency of oscillations in the junction under a voltage. The idea of resonance is basic for most of receivers: a set is "tuned in" when the proper frequency of the receiving contour is adjusted to that of the station. Josephson junction makes a convenient receiving cell. The two advantages are: first, the frequency depends on the voltage and is easily varied; second, the resonance being very sharp results into high selectivity and precision. Josephson elements were employed in the most sensitive detectors for observations of the electromagnetic radiation of the Universe.

23.3 The quantum magnetometer

Josephson effect together with magnetic flux quantization provide a basis for a whole family of supersensitive measuring devices called *SQUIDs.*

This abbreviation stands for *Superconducting Quantum Interference Devices.* We shall tell here about the quantum magnetometer that measures weak magnetic fields.

The simplest quantum magnetometer consists of a superconducting ring with a Josephson junction, Fig. 23.5. As you know, in order to create a current through a normal tunnel contact one must apply some voltage. But for a superconducting junction this is not necessary. Superconducting pairs may tunnel through the insulating layer and superconducting current may circulate in the ring regardless of the Josephson junction. This is called the stationary Josephson effect. (In distinction to the nonstationary Josephson effect accompanied by emission that was described in the previous section.) However the current is limited by a maximal allowable value called the critical current of the junction, I_c. Currents exceeding I_c destroy the superconductivity of the junction and a voltage drop appears across it. The Josephson effect becomes nonstationary.

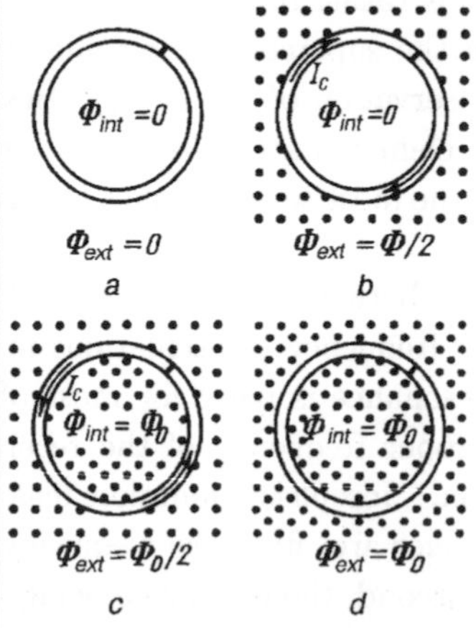

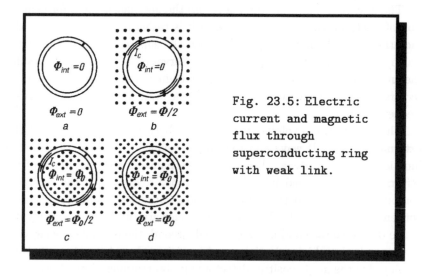

Fig. 23.5: Electric current and magnetic flux through superconducting ring with weak link.

So, insertion of a Josephson junction does not completely destroy superconductivity of the contour. Nevertheless a segment of imperfect superconductivity, the so-called weak link, appears. It plays crucial role in operation of the quantum magnetometer. Let us try to understand this.

In case that the entire contour was superconducting the magnetic flux through it, Φ_{int}, would be strictly constant. Indeed, by the law of electromagnetic induction any change of external magnetic field gives rise to the electromotive force of induction, $\mathcal{E}_i = -\Delta\Phi_{\text{ext}}/\Delta t$, that effects the electric current. The change of the current in its own turn generates the electromotive force of selfinduction, $\mathcal{E}_{si} = -L\,\Delta I/\Delta t$. The resistance of the superconducting contour and the voltage drop in it are zero:

$$\mathcal{E}_i + \mathcal{E}_{si} = 0,$$

and

$$\frac{\Delta\Phi_{\text{ext}}}{\Delta t} + L\frac{\Delta I}{\Delta t} = 0.$$

Remember that the magnetic flux through the contour arising due to the current I is $\Phi_I = L\,I$. This means that $\Delta\Phi_{\text{int}} = \Delta\Phi_{\text{ext}} + \Delta\Phi_I = 0$ and the change of the superconducting current compensates the change of external field. The total magnetic flux through the contour remains constant, $\Phi_{\text{int}} = \Phi_{\text{ext}} + \Phi_I = const$. There is no way to change it without transferring the contour to the normal state. The magnetic flux is "frozen".

What happens if the contour contains a weak link? Then magnetic flux through a contour may change since the weak link allows magnetic quanta to penetrate inside the ring. (You remember that magnetic flux encircled by a superconducting current is quantized and equals a whole number of the quanta Φ_0.)

Let us watch the magnetic flux through a superconducting ring with a weak link and the electric current in it as external magnetic field changes. Let the initial external field and current be zero, Fig. 23.5, *a*. Then the magnetic flux through the ring is zero as well. If we enhance the external magnetic field a superconducting current will arise and the external flux will be completely compensated. This will happen until the electric current reaches the critical value I_c, Fig. 23.5, *b*. To be definite we shall assume that this happens when the flux of the external field equals one half of the quantum: $\Phi_0/2$.[f]

As soon as the value of the current reaches I_c the superconductivity of the weak link is destroyed and the quantum of magnetic flux Φ_0 enters the ring, Fig. 23.5, *c*. The ratio Φ_{int}/Φ_0 stepwise increases by unity. (The

[f]Critical current depends on many factors and in particular on the thickness of the dielectric. It is always possible to fit the latter so that the flux created by the critical current has the desired value: $L\,I_c = \Phi_0/2$. This simplifies analysis but does not affect essentials.

superconducting contour passes to the next quantum state.) And what about the current? The value remains the same but the direction reverses. Judge for yourself, formerly the external flux was compensated by the field of the current: $\Phi_I + \Phi_{\text{ext}} = -L I_c + \Phi_0/2 = 0$. After the quantum has entered the ring the current and the external flux add up: $\Phi_{\text{ext}} + \Phi'_I = \Phi_0/2 + L I'_c = \Phi_0$. Thus letting in the flux quantum instantaneously has changed the direction of the current.

As the external field grows further the current in the ring decreases and the superconductivity of the junction is restored. When the external flux is Φ_0 the current disappears at all, Fig. 23.5, *d*. After that it changes the direction again in order to screen the excess of magnetic flux. Finally, when the external flux comes to $3\Phi_0/2$ the current becomes I_c, the superconductivity of the junction is destroyed and one more quantum of magnetic flux enters the ring, *etc.*

The dependencies of the magnetic flux across the ring, Φ_{int}, and the electric current I versus the external magnetic flux Φ_{ext} are shown in Fig. 23.6. Both fluxes are measured in magnetic quanta Φ_0 which present the natural units. The stepwise shape of the dependence offers a possibility to "count" individual flux quanta despite their extremely small ($\sim 10^{-15}\,Wb$) magnitude. The reason is quite clear. Even though the magnetic flux through the superconducting contour changes by a tiny amount $\Delta\Phi = \Phi_0$ this happens in a very short time Δt, almost instantly. Therefore the velocity $\Delta\Phi/\Delta t$ during this abrupt change may be really big. It can be measured, for example, by electromotive force induced in a special measuring coil of the device. This is the principle of operation of quantum magnetometer.

Construction of a real quantum magnetometer is for sure much more complicated. Say, usually not one but several weak links are connected in parallel. This gives rise to peculiar interference of superconducting currents (or to be exact of the corresponding quantum waves that determine locations of electrons). This helps to increase precision of measurements. The collective name *SQUID* of such devices refers to the interference of quantum waves. The sensitive element of a device is inductively coupled with an oscillatory contour where the jumps of magnetic flow are converted into electric impulses to be amplified later. But these technical subtleties are far beyond the scope of the book.

The fact is that supersensitive magnetometers capable of measuring magnetic fields with $10^{-15}\,T$ accuracy are now widely applied industrial

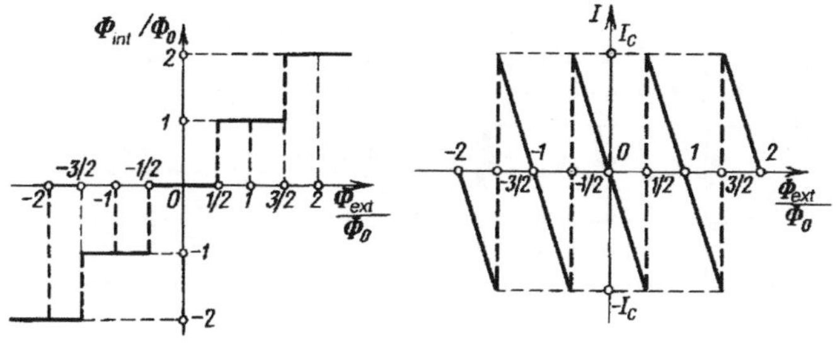

Fig. 23.6: Magnetic flux *(a)* through superconducting ring with weak link and electric current *(b)* in it as a function of external magnetic flux.

production. Among other things they are used in medicine. It turns out that working heart, brain and muscles create weak magnetic fields. For example the magnetic induction due to activity of heart is $B \approx 10^{-11}$ T being a hundred thousand times less than the field of the Earth. But still these fields, whatever weak they are, lie within the reach of SQUIDs. Records of rhythms of these fields are called magnetocardiograms, magnetoencephalograms *etc.* Superconducting facilities offered new possibilities to register and study the most delicate signals of human organism. This was a breakthrough in medical diagnostics of many diseases.

Experiments in the field started in seventies. In order to minimize the influence of the magnetic field of the Earth measurements were carried out in specially designed screened chambers. Their walls were made of three layers of metal with high magnetic permeability that presented efficient magnetic screening plus two layers of aluminum in between for electric screening. These precautions provided the means to reduce magnetic field inside the chamber to several ***nanoTesla*** ($1\,nT = 10^{-9}\,T$) that is tens thousand times less than that of the Earth. Clearly such chambers costed a fortune. Further development of this promising realm of SQUID applications led to remarkable progress and greatly simplified the procedure. Modern superconducting technique permits taking a distinct magnetocardiogram with no screens at all, Fig. 23.7. The only imperative condition is to remove metal clips and the content of your shirt pocket.

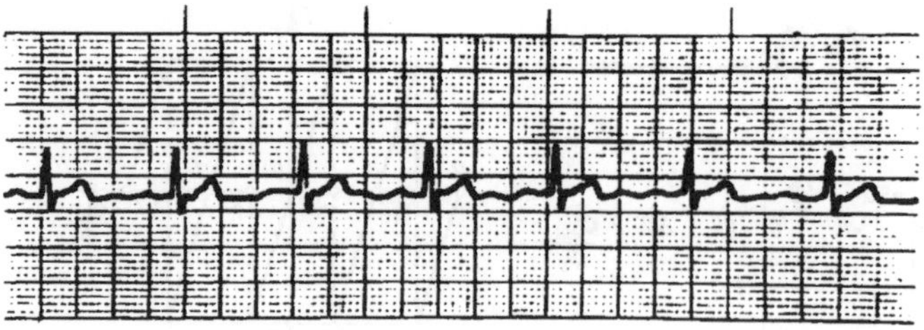

Fig. 23.7: Modern magnetocardiogram.

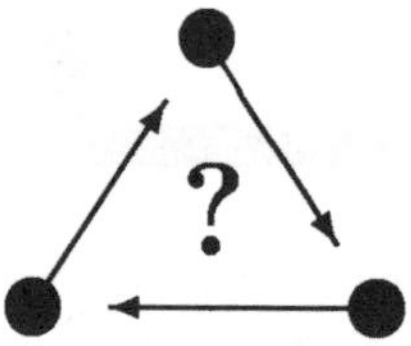

Even the "triple protection" did not eliminate traces of the Earth's magnetic field in magnetocardiographic chambers. What could have been other objectives of building them?

Chapter 24

The superconducting magnets

Strong magnetic fields can be obtained by passing strong electric currents through a coil. The greater is the current the bigger is the field. In case that the coil possesses electric resistance heat is released as the current flows. Supporting the current requires enormous energy and, besides, a serious problem is to carry away the heat which may fuse the coil. In 1937 one has first realized a magnetic field with the induction $10\,T$ this way. But the field could be supported only at night when all other consumers were disconnected from the power station. The liberated heat was removed by running water and 5 liters (1.3 *gal*) of it were brought to a boil every second. The heat release sets the main limitation to creating strong magnetic fields by ordinary coils.

As soon as superconductivity was discovered the idea appeared to exploit it in production of strong magnetic fields. At the first sight the only thing to be done is to wind up a coil of superconducting wire, send around a strong enough current and short the circuit. Once the resistance of the coil is zero no heat is released. The gains would justify the work done when cooling the solenoid down to the temperature of liquid helium unless... magnetic field destroyed superconductivity.

The way out was found. The help came from laws of quantum mechanics. As you know, in superconductivity those may work on macroscopic scales.

24.1 The Meißner effect in detail

In Fig. 24.1 you may see the scheme of the experiment that was performed by Kamerlingh Onnes in 1911 in Leiden. The Dutch scientist put a lead coil into liquid helium where it cooled down to the helium boiling point. The electric resistance of the coil disappeared because it turned into the superconducting state. After that he reconnected the switch and closed the coil onto itself. The undamped superconducting began to circulate in the coil.

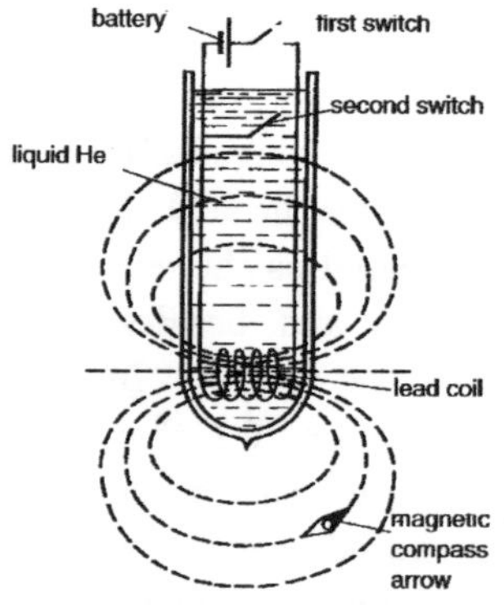

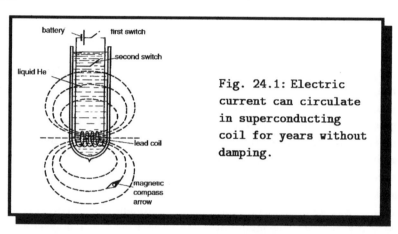

Fig. 24.1: Electric current can circulate in superconducting coil for years without damping.

The current generates magnetic field with the induction proportional to its strength. A naive assumption is that the larger is the current in the coil the bigger magnetic field it produces. But the results were discouraging: as the field reached several hundredths of *Tesla* the solenoid passed to the normal state and electric resistance appeared. Attempts were done to prepare coils of other superconductors but in those again superconductivity was destroyed at relatively weak fields. What was the rub?

The puzzle of such "inconvenient" behavior of superconductors was solved in 1933 in the laboratory of W. Meißner in Berlin. It was found that superconductors possess the property of expelling magnetic field; the induction inside superconductors is zero. Imagine that a metal cylinder (a piece of wire) was cooled and became superconducting. Then one switched on a magnetic field with the induction $\vec{B}_{\text{ext}}$. By the law of electromagnetic induction this must cause at the surface of the cylinder circular currents, Fig. 24.2. The magnetic field $\vec{B}_{\text{cur}}$ created by the currents inside the cylinder is equal to $\vec{B}_{\text{ext}}$ in magnitude but opposite in direction. The currents are superconducting and do not die out. Therefore the net induction in superconductor is zero: $\vec{B} = \vec{B}_{\text{ext}} + \vec{B}_{\text{cur}} = 0$. Lines of magnetic induction do not penetrate superconductors.

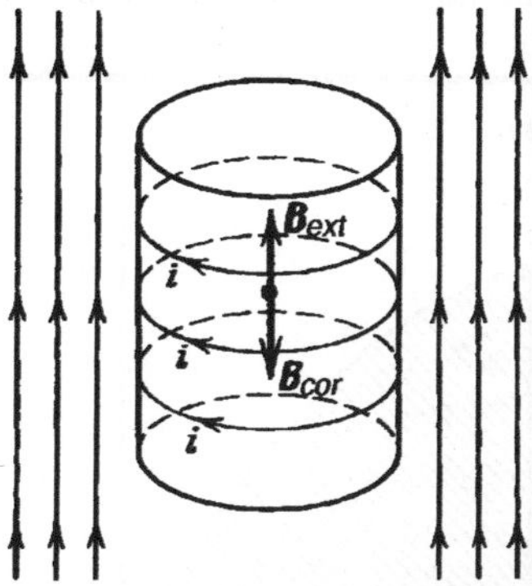

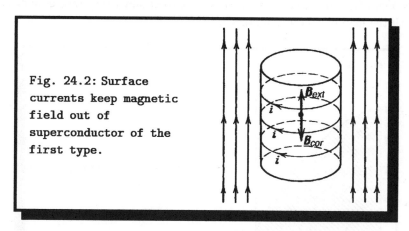

Fig. 24.2: Surface currents keep magnetic field out of superconductor of the first type.

But what if we change the order and apply the field before cooling the specimen to superconducting state? It seems that the magnetic induction will not change and there will be no point in generating surface currents. This was the logic of Meißner when he checked calculations by Laue[a] concerning the first experimental procedure. But still he preferred to check. The result of the renewed experiment was stunning. It turned out that magnetic field was just the same forced out of superconductor without penetrating it. This was called the Meißner effect.

Now it is clear why magnetic field destroys superconductivity. Exciting surface currents takes energy. In this sense superconducting state is less favorable than normal one when magnetic field enters the bulk and there are no surface currents. The higher is the induction of external field the stronger screening current it demands. At some value of magnetic induction the superconductivity inevitably will be destroyed and the metal will transform to normal state. The value of the field when the destruction of superconductivity occurs is called the critical field of the superconductor. It is important that presence of external field is not a necessary condition of the destruction. Electric current in the superconductor produces a

[a]M. von. Laue, (1879–1960), German physicist; Nobel Prize 1914.

magnetic field of its own. When at certain intensity of the current the induction of the field reaches the critical value the superconductivity breaks down. The value of critical field increases at low temperatures but even near the absolute zero critical fields of pure superconductors are modest, see Fig. 24.3. So it could seem a vain hope to obtain strong magnetic fields with the help of superconductors.

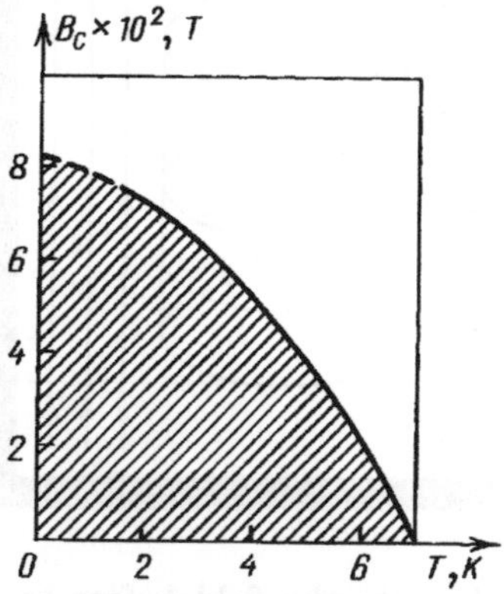

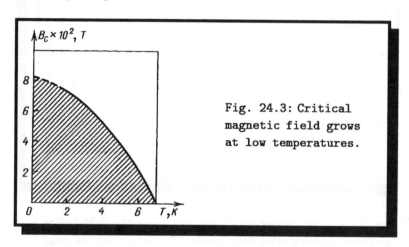

Fig. 24.3: Critical magnetic field grows at low temperatures.

But further investigations in the field proved that the situation is not desperate. It was found that there is a whole group of materials that stay superconducting even in very strong magnetic fields.

24.2 The Abrikosov vortices

As it was already mentioned above, in 1957 the prominent theoretical physicist A. A. Abrikosov[b] showed that magnetic field does not destroy superconductivity of alloys so easily. Similarly to the pure case magnetic field begins penetrating into superconductor at some critical value of induction. But in alloys the field does not occupy the entire volume of the superconductor at once. At first only detached bundles of magnetic lines are formed in the bulk, Fig. 24.4. Every bundle carries an exactly fixed portion. It is equal to the quantum of magnetic flux, $\Phi_0 = 2 \cdot 10^{-15}\, Wb$, that we have

[b]A. A. Abrikosov, (born 1928), Russian physicist, pupil of L. D. Landau, specialist in condensed matter physics.

already met[c].

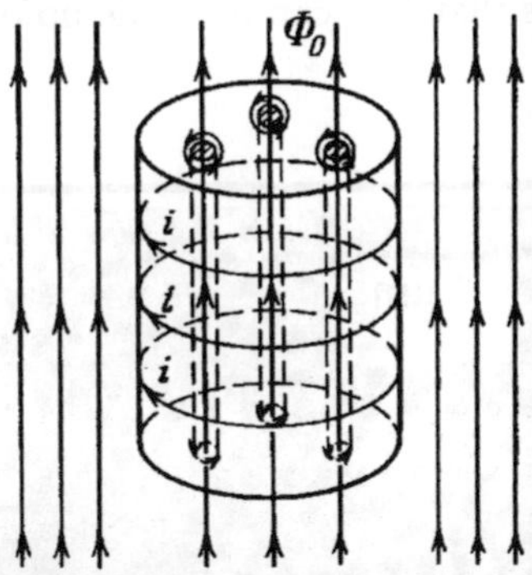

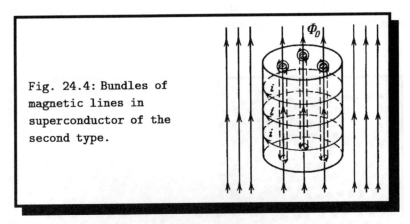

Fig. 24.4: Bundles of magnetic lines in superconductor of the second type.

The stronger is the magnetic field the more bundles enter the superconductor. Each of them brings one magnetic quantum and the total flux changes stepwise. Again, like before, magnetic flux through superconductor may take only discrete values. It is astonishing to see the laws of quantum mechanics "working" on macroscopic scales.

Each bundle of magnetic lines piercing the superconductor is enveloped by undamped circular currents that resemble a vortex in gas or liquid, Fig. 24.4. For this reason the bundles of magnetic lines together with the superconducting currents around it are called Abrikosov vortices. Certainly in the core of the vortex the superconductivity is broken. But in the space between the vortices it is conserved! Only in very strong fields when numerous vortices begin overlapping the superconductivity is destroyed completely.

This remarkable reaction of superconducting alloys to magnetic fields was first discovered "at the tip of the pen". But modern experimental technique makes it possible to observe Abrikosov vortices directly. Fine magnetic powder is applied to the surface of superconductor (for example, to the base of a cylinder). The particles gather at the places where the field enters the alloy. Electron microscope study of the surface reveals the dark spots.

Such a photograph of the structure of Abrikosov vortices is shown in

[c]It would be quite natural to say that each magnetic quantum corresponds to one line of magnetic induction. —*A. A.*

Fig. 24.5. We notice that the vortices are arranged periodically and form a pattern similar to a crystal lattice . The vortex lattice is triangular (this means that it is can be made up of periodically repeated triangles).

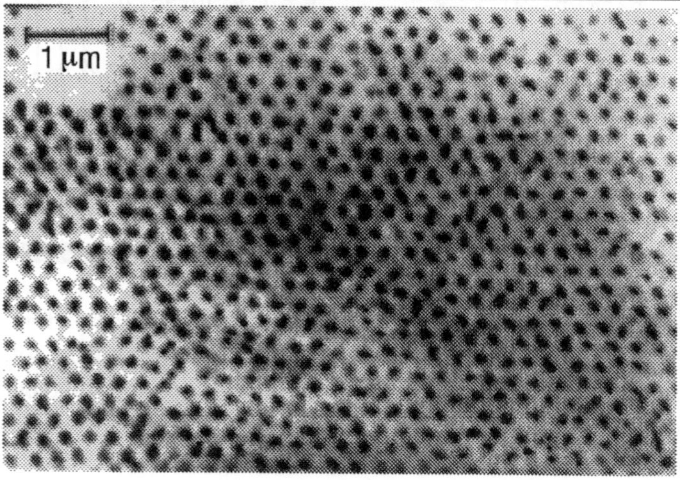

Fig. 24.5: Experimental observation of vortex lattice.

So, in distinction to pure metals alloys, possess not one but two critical fields: the lower critical field marks the moment when the first vortex enters the superconductor and the upper critical field corresponds to the completely destruction of superconductivity. Over the interval between these two the superconductor is pierced by vortex lines. This is called the *mixed state*. Superconductors exhibiting such properties are now called the second type ones. The first type refers to those where the magnetic destruction of superconductivity happens at once, abruptly.

It looked that the problem of producing superconducting magnets was solved. But nature had kept for researchers one more catch. The wire for superconducting solenoids must withstand not only strong magnetic field but strong electric current as well. This happened to be not the same.

24.3 What is pinning?

It is well known that a force acts onto electric current in magnetic field. But where is applied the counteraction force that must appear by the third Newton law? When the field is due to another current then, no doubt, that experiences an equal in magnitude and opposite in direction force (interaction of energized conductors obeys the Ampère[d] law). Our case is more exquisite.

A current that flows in a mixed state superconductor interacts with the magnetic field in the cores of vortices. This affects the distribution of current but the domains where the magnetic field concentrates do not remain intact either. They start moving. Electric current compels Abrikosov vortices to move!

The force exerted onto a current by magnetic field is perpendicular to the magnetic induction and to the conductor. The force acting onto Abrikosov vortices is also perpendicular to the induction of the field and to the direction of the current. Suppose that a current traverses the superconductor depicted in Fig. 24.5 from left to right. Then the Abrikosov vortices will move either up- or downwards depending on the direction of the magnetic field. However the transport of the Abrikosov vortex across superconductor is the motion of the normal non-superconducting core. It suffers a sort of friction which brings on heat evolution. Current in the mixed state superconductor just the same meets a resistance. It could look like these materials were no good for solenoids.

What is the solution? It is to block the motion and hold the vortices in place. Fortunately this is possible. One has simply to worsen the superconductor by making defects in it. Usually defects appear by themselves as a result of mechanical or thermal treatment. Fig. 24.6 demonstrates an electron microscope photograph of niobium nitride film. The critical temperature of the film is 15 K. It was obtained by means of sputtering metal onto glass plate. One clearly discerns the granular (or rather columnar) structure of the material. It is not so easy for a vortex to jump the boundary of a grain. Hence up to a certain current strength, the so-called critical current, vortices stay in place and the electrical resistance is zero.

This phenomenon is known as ***pinning,*** because of vortices being pinned

[d] A. M. Ampère, (1775–1836), French physicist, one of the foundators of classical electrodynamics.

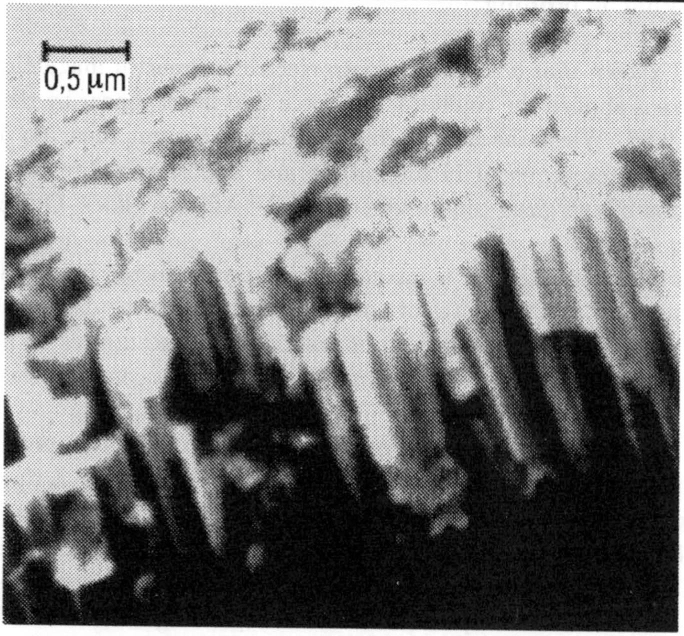

Fig. 24.6: Microscopic structure of niobium nitride film.

by defects.

Pinning offers a possibility to prepare superconducting materials exhibiting high critical values of both magnetic field and electric current. (It would be more accurate to speak not of critical current but of critical current density, that is the current crossing a unit area of cross section.) Critical field is determined by properties of material. In the mean time critical current depends on methods used in preparation and treatment of conductor. Modern technology provides a means to obtain superconductors with high values of all critical parameters. For example, starting from the tin-niobium alloy one can fabricate a material with the density of critical current reaching several hundreds amp/cm^2, the upper critical field equal to $25\,T$ and the critical temperature being $18\,K$.

But this is not the end of the story. It is important whether mechanical properties of the material permit to make a coil. The tin-niobium alloy by itself is too fragile and it would be impossible to bend such a wire. So the following procedure was invented: a copper tube was stuffed with a mixture of niobium and tin powders, then the tube was stretched into a wire and the coil was wound and at last heating the coil made the powders fuse. This resulted into a solenoid of the $Nb_3\,Sn$ alloy.

Industry prefers more practical materials such as the more plastic niobium-titanium alloy Nb Ti. It is used as a base for so-called composite superconductors.

First one drills in a copper bar a number of parallel channels and inserts there superconducting rods. Then the bar is stretched into a long wire. The wire is cut and the pieces inserted into another drilled copper bar. That is once more stretched, cut into pieces and so on... Finally one obtains a cable that contains up to a million of superconducting lines, like those shown in Fig. 24.7. This is used for winding coils.

The important advantage of such cables is that the superconducting current is distributed among all the lines. When compared to superconductor copper behaves like an insulator. If copper and superconductor are connected in parallel then the entire current will choose the path that has no resistance. There is the second advantage. Suppose that by accident superconductivity breaks down in one of the lines. This causes heat liberation and the danger that the whole cable will pass to the normal state. It is urgent is to remove the heat. Copper is a good heat conductor and perfectly suits the purpose of thermal stabilization. Besides it secures good mechanical properties of cables.

***Postscriptum* for taxpayers**

After having started with the high-temperature thriller we turned to applications of conventional superconductors. Here, in contrast to high-temperature ones, the physics of the phenomenon is clear. Nevertheless the lack of theoretical understanding does not stop search for practical applications of high-temperature superconductors. The main stumbling block are bad technological properties of available high-temperature superconductors: they are extremely brittle and do not stand rolling which is an essential element of mechanical treatment of metals. Nevertheless several brands of some kilometers long high-temperature superconducting cables are already on the market. They are produced by rolling and annealing of

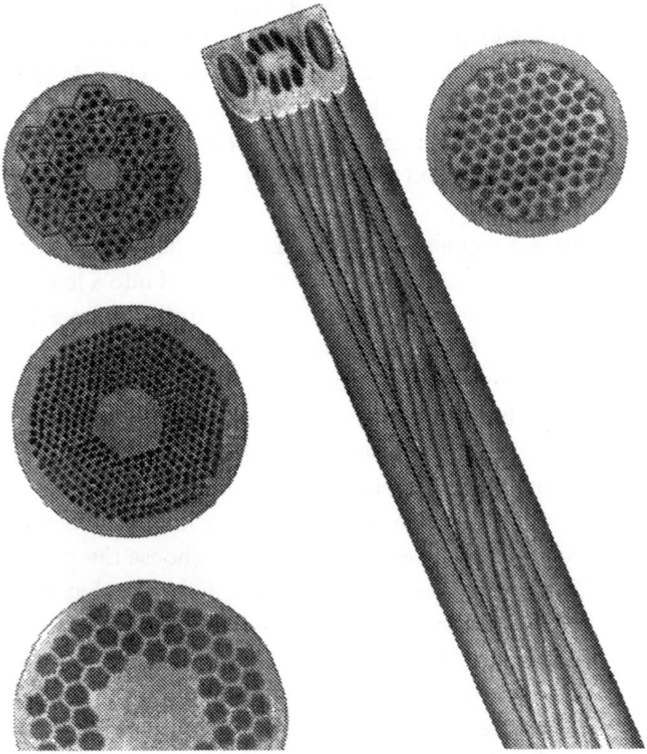

Fig. 24.7: Industrial cables are made of copper with millions of superconducting lines in it.

a tube of silver or other suitable metal filled by high-temperature superconductor powder. A number of experimental underground transmission lines made of such cables are in operation now in France and in USA. The first electric motors and generators based on high-temperature superconductors are under testing. There is no doubt that the field of applications of these materials will expand and new more practical high-temperature superconductors will appear.

Let us turn to prospects. Those are really fantastic. Many of global projects of the past are put back onto agenda because the advent of high-temperature superconductivity makes them profitable. For example, at present 20–30% of all produced electrical energy is wasted in power transmission lines. Using high-temperature superconductors for energy transmission could eliminate these losses.

All projects involving thermonuclear synthesis need giant superconducting magnets that keep high-temperature plasma away from the walls of the chamber. Streams, if not rivers, of liquid helium are necessary to maintain the superconducting state. The helium would be replaced by nitrogen at a tremendous cost saving.

Gigantic superconducting coils would serve as accumulators of electrical power, which would share the load during peak periods.

Supersensitive equipment for making magnetocardiograms and magnetoencephalograms, based on the use of superconducting Josephson elements, would come to every hospital.

Magnetic cushions created by superconducting coils would support intercity express trains commuting at speeds of $400 - 500\ km/h$.

A new generation of supercomputers based on superconducting elements and cooled by liquid nitrogen would be constructed.

Don't think we've lost our heads over high-temperature superconductivity. Since its discovery, the ardor of many investigators has notably cooled down. The same happens when an Olympic record stays out of reach for years. But as soon as the record has been set it serves a benchmark. The possibility of producing materials with unique characteristics has been confirmed. Certainly not once economic considerations will affect realization of projects and it is not tomorrow that we will surpass the records and make them a routine. But today we know for sure that the impossible has become accessible. And this has irreversibly changed the reference point in our attitude toward superconductivity.

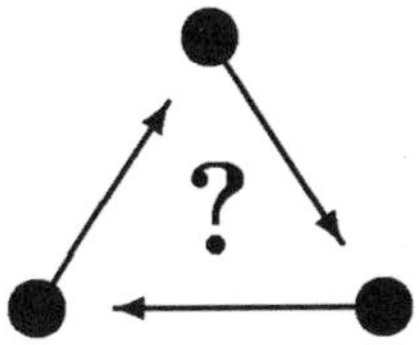

Why superconducting transmission lines do not require expensive high-voltage equipment?

Afterword

Little by little our tale about physics came to the end. We told you how physics helps to explain so many things all around us. Remember meandering rivers and the blue sky, think of coalescing droplets and hissing tea-kettles, don't forget the singing violin and the chime of goblets. Still the magic of physics is not solely the power to explain what happens but the ability to foresee what will happen even if it never has before. This gained physics the head position in scientific and technical progress of our days.

Modern physics has opened to us the amazing quantum world. There prisoners of potential wells flee away from their dungeons like the Count of Monte Cristo; magnetic fields make vortices to pierce superconductors; volatile amalgam of wave and particle entities of light quanta brings to mind mythical centaurs. Wonders of the quantum world are beyond imagination. But using its mathematical arsenal theoretical physics succeeds to describe behavior of quanta so accurately that results of experiments exactly coincide with theoretical predictions. This capability to correctly represent phenomena which escape even mental visualization was, in the opinion of the world-known physicist L. D. Landau, the greatest triumph of theoretical physics of twentieth century.